Ethereum: The Platform Fueling the Smart Contract Revolution

Sana

Contents

5 Evaluation   43

6 Conclusions   59

A Source code   61

B Bug analysis

Chapter 1

Introduction

1.1 Motivation

The success of BitCoin [36], the first decentralised cryptocurrency, has raised considerable interest both in industry and in academia. Cryptocurrencies feature a distributed protocol where a set of nodes maintain and agree on the state of a distributed public ledger called blockchain. Although Bitcoin is the most paradigmatic application of blockchain technology, there are other applications beyond cryptocurrencies, such as financial, digital identity verification, voting and even government services [50, 60, 63].

To enable these general-purpose applications, blockchains allow the deployment of smart contracts that can act as autonomous agents to govern agreements between mutually distrusting participants. The most prominent blockchain platform for smart contracts is Ethereum [4, 20, 61]. Ethereum allows the deployment of contracts, written in languages such as Solidity [52] and Vyper [59], in the form of Ethereum Virtual Machine (EVM) bytecode. A contract's EVM bytecode and state are stored in the blockchain, and the contract executes within blockchain transactions, possibly manipulating Ethereum currency, called ether, in their execution.

Although smart contracts are promising to drive a new wave of innovation, there are a number of challenges to be tackled. Being a fairly new technology that governs a growing universe of valuable assets, smart contracts are prone to errors and in particular malicious attacks. In fact, Ethereum already faced several devastating attacks on vulnerable smart contracts, like the DAO hack in 2016 [51] and the Parity Wallet hack in 2017 [41], together causing an estimated loss of over 400 million US dollars. These attacks were related with bad coding practices and unforeseen consequences of the code implementation and decisions. Moreover, a new trend is that attackers try to lure their victims into traps by deploying seemingly vulnerable contracts that contain hidden traps [54], instead of searching for vulnerable contracts anymore.

Beyond specific security issues and malicious attacks, Ethereum contracts can be in general unreliable in their implementation due to a number of reasons. Solidity is the most widely used language for such contracts, yet it is still in alpha stage (version 0.7 as of September 2020),

and "breaking changes as well as new features and bug fixes a re i ntroduced r egularly" [52].
Many vulnerabilities seem to be caused by a misalignment between the semantics of Solidity and
the intuition of programmers and the specific context of a b lockchain. T he l anguage d oes not
introduce constructs to deal with domain-specific aspects, like t he fact t hat computation steps
are recorded on a public blockchain, wherein they can be unpredictably reordered or delayed [2],
the persistence of smart contracts in the blockchain (once deployed, smart contracts cannot be
modified o r r emoved u nless d uly p rovisioned w ith s uch m echanisms), o r t he i nteraction with
with other smart contracts and external off-chain services.

1.2   Problem statement

As with most software, the verification of s mart c ontracts c an e mploy t echniques f rom t he realm
of static analysis, to scan and look for potential bugs and vulnerabilities, as well as dynamic
analysis, that executes code with the same purpose. We survey the main approaches later in this
book. The most commonplace assurance for smart contracts is provided by unit testing, a
sometimes limited yet many times the most practical form of dynamic analysis: a set of test cases
is defined to exercise a contract's functionality, where each test case is defined through a set of
fixed inputs, a call sequence that exercises the software, and test assertions that match the observed
behavior versus the expected one. Unit testing can be ineffective in finding bugs, as the choice of
inputs and exercised behaviours is limited to what the test programmer could think of as suitable/
reasonable and could in practice program. Thus, bugs due to edge cases and/or less common
interactions can easily be missed.

Property-based testing (PBT) is an approach that can overcome these drawbacks. It works
by automatically deriving test cases and their inputs according to a model of correctness for the
software under test, seeking falsifying examples for the violated model as witnesses of deviant
behavior, and then, in a process known as shrinking, reducing their complexity or length towards
edge cases that facilitate human understanding. Thus, a programmer may focus on specifying the
properties of interest for the software and input generation constraints through a model, rather
than making a specific choice of i nputs, a nd a n a rbitrary n umber of t est c ases m ay potentially
be generated at random in a model-driven manner.

We apply the PBT methodology for verifying a highly important type of smart contracts,
ERC-20 tokens. The ERC-20 specification [58] u nderlies m ost c ontracts i n t he a rea o f token
management. It allows the uniform management of custom tokens enabling decentralised exchange,
in particular of most digital coins that work on top of Ethereum [23], and empowers distributed
applications called Dapps [33] that interface with smart contracts, e.g., digital wallets. Our
proposal provides a methodology for verifying such contracts during development or finding bugs
in real-world ERC-20 contracts that have already been deployed to the Ethereum blockchain.

1.3 Contributions

The overall contribution of this book is a PBT framework for ERC-20 contracts, and its evaluation using real-world contracts. In more detail:

- The ERC-20 testing framework takes form through a rule-based state machine model deployed on top of Brownie [3], a Python-based development and testing framework for Ethereum smart contracts that incorporates the PBT Hypothesis engine [28, 32]. Since the model is a general one, any ERC-20 contract can be tested at will. Moreover, it is extensible, as we illustrate for a few other extra functionalities that are common in contracts: token minting, burning, and sale (exchange by ether).

- The evaluation of this approach covers 10 contracts written in Solidity, including the two reference implementations and eight real world-examples, some of which are widely used ones. The evaluation covers bug findings, a detailed performance analysis, a comparative assessment to bug findings by the ERC-20 reference implementation's unit testing suites, and, finally, also results for ERC-20 extended functionality.

- The source code for the software developed in the scope of this book, the contracts used for evaluation, and the unit testing evaluation frameworks were made available at GitHub [45–47].

1.4 book structure

The rest of this book is structured as follows:

- Chapter 2 puts forward the background concepts underlying this work, and related work in the state-of-the-art testing tools and methods relevant to this work.

- Chapter 3 presents our PBT framework, with a prior overview of ERC-20 contracts and the Brownie framework.

- Chapter 4 details the design and implementation of the testing framework.

- Chapter 5 describes and displays the results of the conducted evaluation.

- Chapter 6 ends with a summary of the main conclusions of this book, and highlights directions for future work.

The text of the book is supplemented by two appendices:

- Appendix A lists the main source code for the PBT framework.

- Appendix B contains a detailed analysis of ERC-20 bug findings for the contracts evaluated.

Chapter 2

Background

This chapter introduces fundamental concepts to provide a better understanding of the subject covered in this book. We start introducing the necessary concepts and background necessary to understand smart contracts: the general notion of blockchain (Section 2.1); a description of the Ethereum blockchain (2.2); and Ethereum smart contracts, their execution, and ERC-20 tokens (2.3). We address the functionality of ERC-20 contracts in detail only in Chapter 3. We then describe the property-based testing approach (2.4), and finish the chapter with a discussion of related work to this book (2.5).

2.1   Blockchain

A blockchain is an append-only data structure made of data blocks, where each block comprises multiple transactions or digital events, as illustrated in Figure 2.1. On top of this data structure, a distributed public ledger keeps track of each transaction that took place since the genesis block (the first block) until the present. In addition to the transactions, each block contains a timestamp, the hash value of the previous block, and a nonce, which is a random number for verifying the hash [64]. The blockchain is stored, maintained, and collaboratively managed by a distributed group of participants called nodes. It is resilient to attempts to corruption or spoofing of existing blocks by cryptographic mechanisms and consensus protocols [62] that work through Proof-of-Work mechanisms in Bitcoin and Ethereum or Proof-of-Stake in Ethereum 2.0 [21] or other blockchains like Tezos [25].

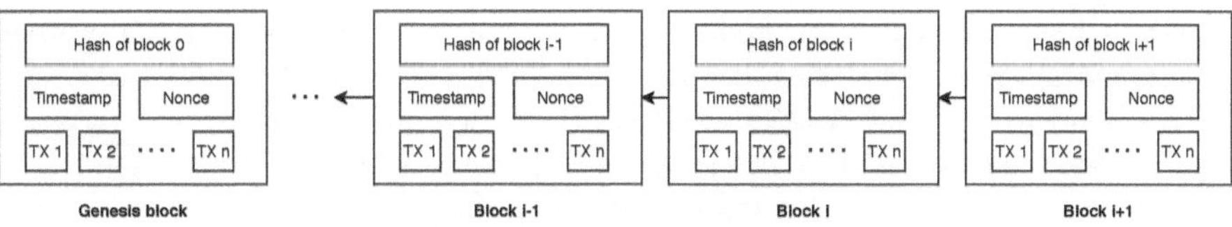

Figure 2.1: A Blockchain data structure

Bitcoin [36] is the most popular example of blockchain technology. The digital currency Bitcoin itself is highly controversial but the underlying blockchain technology is quite robust. Bitcoin has been employed in a wide range of applications in both financial and non-financial world [8]. An example is Namecoin [37], a decentralized name registration database based on the Bitcoin technology, which can register and provide Human-readable Tor .onion domains or even decentralised TLS certificate validation backed by blockchain consensus.

2.2   Ethereum

Ethereum [4] is a global, open-source platform for decentralized applications. It is a blockchain with built-in support for Turing-complete programming languages, which allow anyone to write smart contracts and decentralised applications where they can create their own arbitrary rules for ownership, transaction formats and state transition functions. Ethereum's vision is to create a censorship-resistant self-sustaining decentralised world network, where autonomous user-defined programs called smart contracts can specify rules for governing transactions, thus removing the need for a central governance entity and enforced by a network of peers. A tradable cryptocurrency, called Ether, is built-in in the blockchain, and is used by users and contracts to pay for transaction fees and services on the Ethereum network.

The Ethereum Virtual Machine (EVM) [61] is a sandboxed virtual stack environment that runs within each Ethereum node. The execution of contracts, expressed using the EVM bytecode format, is completely isolated from the network, filesystem or any processes of a local node. To counter for computationally expensive operations and potential denial-of-service attacks, every opcode has its own base gas cost, and transactions are bound by a gas limit. When a user wants to initiate a transaction, they reserve some Ether which they are willing to pay for the gas cost associated with the execution of that particular transaction. Thus, EVM implements a payable scheme that charges per software instruction executed instead of per financial transaction executed, like Bitcoin does.

2.3   Smart contracts and ERC-20

Ethereum supports two kinds of accounts, user and contract accounts. Both can have balance, be owned by an Ethereum address, and publicly reside on the blockchain. In contrast to a user account, a contract account is an autonomous agent managed by its own code. The contract code captures agreements between mutually distrusting parties, which are enforced by the consensus mechanism of the blockchain without relying on a trusted authority. Contracts also have persistent state where the code may store data, such as token balances, auction bids or anything else that represents a digital asset. Smart contracts are thus formed by their accounts, code, and persistent state.

Smart contracts are written in high-level languages such as Solidity [52] and Vyper [59] that are compiled to EVM bytecode. The EVM bytecode for a function contract is executed whenever it receives a corresponding invocation message, either from a user or from another contract. During execution, a contract may read from or write to its storage file, receive Ether into its account balance, and send Ether to other contracts or users. Conceptually, one can think of a contract as a special "trusted third party" in terms of availability and correctness of execution in terms of the EVM. A contract's entire state is visible to the public, as well as its complete transaction history, ensuring non-repudiation and allowing transparent audits to their functionality (which may still be unreliable or insecure).

Given that there are no strict rules about how smart contracts should behave, the Ethereum Foundation community has developed a variety of standards and guidelines for how a contract should behave and inter-operate with other contracts. The standards for smart contracts are called ERCs (Ethereum Request for Comments), a branch of the more general EIPs (Ethereum Improvement Proposals). ERC-20 [58], the focus of this book, is the most well-known and widely implemented standard for managing tokens. Tokens represent blockchain-based assets of value, and their operation is governed by smart contracts in terms of creation, destruction, or exchange. Tokens can represent fungible assets like money, time, or shares in a company, but also non-fungible ones such as domain names [37] or virtual pets [10].

2.4 Property-based testing

Property-based testing (PBT) is a methodology for software testing that first became popular with the QuickCheck library for Haskell [6], and also implemented later in Erlang [1]. Similar libraries exists for other programming languages where PBT is also quite popular, for instance ScalaCheck for Scala and Java [38], or Hypothesis for Python [32] (used in this work). PBT is used for testing general-purpose software, as well as domain-specific applications in diverse fields, e.g., compilers [43], theorem provers [13], stream processing [44], robotic platforms [48], or telecommunications software [1].

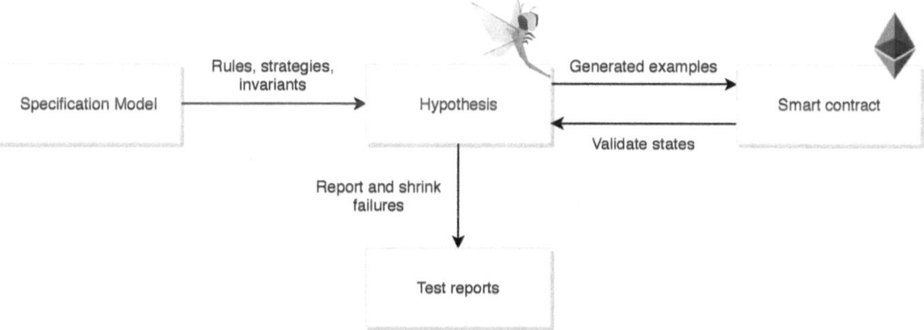

Figure 2.2: Property-based testing approach.

The overall PBT approach we take in this book is illustrated by Figure 2.2. We use the Hypothesis framework for PBT, integrated in the Brownie environment for the development of Ethereum smart contracts [3], to generate test cases for smart contracts. PBT works by automatically deriving test cases and random inputs according to a model of correctness for the software under test, seeking falsifying examples for the violated model as witnesses of deviant behavior, and then, in a process known as shrinking, reducing their complexity or length towards edge cases that facilitate human understanding.

Using PBT, a programmer may thus focus on specifying the properties of interest for the software and input generation constraints through a model, rather than making a specific choice of inputs, and an arbitrary number of test cases may potentially be generated at random in a model-driven manner. At the same time, the use of larger set of inputs potentially extends test coverage to include edge cases or complex program interactions that may easily be missed by a programmer when coding unit tests.

PBT is similar to fuzz-testing in the sense of using randomisation for the generation of inputs. Raw fuzz-testing techniques, however, tend to employ low-level techniques for the generation of input values and to look for "extreme" program behaviours that may be significant in terms of security like program crashes, information leaks, etc. In PBT, by contrast, the process is model-driven and the aim of testing is to verify user-defined functional properties.

2.5   Related work

Several approaches have been taken to verify Ethereum smart contracts and detect potential issues of correctness and/or security. A recent survey is provided in [14], where testing tools can either be based on static or dynamic analysis.

Static analysis works by scanning a contract's source code or EVM bytecode for security vulnerabilities and bad coding practices without executing the contract. An example is Securify [49, 56], which derives semantic facts inferred by analysing the contract's dependency graph and uses these facts to check a set of compliance and violation patterns. Based on the outcome of these checks, it classifies all contract behaviours into violations, warnings, and compliant. SmartCheck [53] is another static analysis tool, which flags potential vulnerabilities in Solidity contracts by searching for specific syntactic patterns in the source code. It works by translating the source code into an XML-based intermediate representation and then checks the intermediate representation against XPath patterns to identify potential security issues.

Dynamic analysis attempts to check how the code behaves during execution to see if changes resulted in invalid states. An example is Oyente [31], one of the first smart contract analysis tools that uses symbolic execution on EVM bytecode to identify vulnerabilities. It executes EVM bytecode symbolically and checks for deviant execution traces for a number of possible cases: transaction order influences Ether flow, the result of a computation depends on the timestamp of the block, exceptions raised by calls are not properly caught, or a contract is re-entered multiple times [14]. Oyente served as a starting point for several other projects and is considered to be a reference tool. Another example is Mythril [35]. Developed by ConsenSys, it relies on symbolic analysis, taint analysis and control flow checking of the EVM bytecode to prune the search space and to look for values that allow exploiting vulnerabilities in the smart contract. That is, it executes EVM bytecode symbolically by constructing a control flow graph, where nodes contain disassembled code and edges are labeled with path formulas. Mythril is considered to be the most accurate tool for detecting vulnerable smart contracts as according to a survey where from a dataset of annotated vulnerable smart contracts, Mythrill was able to detect 27% of the vulnerabilities [15].

On the other hand, recent work surveying and categorising flaws in critical contracts established that fuzzing using custom user-defined properties might detect up 63% of the most severe and exploitable flaws in contracts [27]. One example is the Echidna [16], an open-source smart contract fuzzer. Echidna generates tests to detect violations in assertions and custom properties. It makes use of user-defined properties (property-based testing), assertion checking, and gas use estimation instead of relying on a fixed set of pre-defined bug oracles to detect vulnerabilities. It also offers responsive feedback, captures many property violations, and its default settings are calibrated based on experimental data [26].

Chapter 3

Property-based testing of ERC-20 contracts

This chapter exposes our PBT approach to ERC-20 contracts. We first provide an overview of how ERC-20 contracts (should) work, are implemented in Solidity, and examples of real-world bugs (Section 3.1), Next, we describe the base support for PBT of Ethereum smart contracts provided by the Brownie framework (3.2). We then present our PBT approach through the definition of rule-based state-machines implemented on top of Brownie for the ERC-20 model and a few other common extensions to ERC-20 found in contracts (3.3). Finally, the use of the framework is illustrated in terms of test instantiation, execution, and examples of bug detection (3.4).

3.1 ERC-20 contracts

3.1.1 The ERC-20 specification

ERC-20 [58] defines a standard interface for the creation of tokens on the Ethereum blockchain. A contract maintains a total supply of tokens that are owned in association to accounts identified by addresses in the blockchain. Tokens can be transferred by the owners to other recipient addresses, and a complementary mechanism of allowances permits third-parties to perform a transfer on behalf of the owner. These functionalities require state to be maintained for the contract in terms of total supply, account balances, and allowances, plus events to be recorded for transactions that correspond to ERC-20 function invocations.

In correspondence to this overall functionality, ERC-20 defines precise operations and their expected behavior for compliant contracts. The operations are declared in the Solidity interface shown in Listing 3.1. A Solidity interface only lists public properties and function signatures[1],

---

[1] In newer versions of the language, the `interface` keyword explicitly denotes an interface. The code here uses the more general `contract` declaration.

Listing 3.1: EIP20 interface

```
1  contract EIP20Interface {
2    uint256 public totalSupply;
3    function balanceOf(address owner)
4      public view returns (uint256 balance);
5    function transfer(address to, uint256 value)
6      public returns (bool success);
7    function allowance(address owner, address spender)
8      public view returns (uint256 remaining);
9    function approve(address spender, uint256 value)
10     public returns (bool success);
11   function transferFrom(address from, address to, uint256 value)
12     public returns (bool success);
13   event Transfer(address indexed from, address indexed to, uint256 value);
14
15   event Approval(address indexed owner, address indexed spender,
16                  uint256 value);
17 }
```

and the code shown is taken from the Consensys reference implementation of ERC-20 [7]. The operations and their expected behavior are as follows:

- The totalSupply() function (line 2 in Listing 3.1) returns the total supply of tokens, a 256-bit unsigned number as expressed by the uint256 type in Solidity[2].

- A call to balanceOf(owner) (line 3) returns the balance associated to account with address owner.

- A call to transfer(to, value) (line 5) transfers value amount of tokens from the caller's implicitly defined address, denoted by msg.sender in Solidity, to address to. The operation is allowed if balanceOf(msg.sender) >= value, and in that case:

  - the balance of msg.sender and to must be updated, i.e., respectively decremented and incremented by amount;

  - the Transfer(msg.sender, to, value) event must be emitted for the transaction (this type of event is defined at line 13);

  - and the method must finally return true.

Otherwise, the implementation should revert the transaction by throwing an exception. This means that an error is signalled, but also that any changes made so far to the contract state are not committed to the blockchain. The specification also implicitly allows a false return value in place of the exception throwing (a more robust approach). This introduces

---

[2]The code defines a the totalSupply public attribute and an implicitly-defined "getter" function with the same name.

some ambiguity: as we illustrate later in this chapter and Chapter 5, some contracts return
false instead of reverting the transaction, while others omit a return type for transfer
altogether given the revert mechanism.

- The remaining functions are related to allowances and corresponding transfer operations:

  - A call to allowances(owner, spender) (line 7) returns the current allowance of spender
    for tokens owned by owner, i.e., how many tokens spender is allowed to transfer on
    behalf of owner through the transferFrom function discussed below.

  - A call to approve(spender, value) (line 9) sets an allowance of value tokens owned by
    msg.sender for spender. If successful, the function must emit an Approval(msg.sender,
    spender, value) event (this type of event is defined at line 15) and return true.

  - A call to transferFrom(from, to, value) is used by the caller to make a transfer of
    value tokens between accounts from and to. The function works similarly to transfer,
    but has the additional pre-condition that allowance(owner, msg.spender) $>=$ value and
    the additional post-condition of decrementing the allowance at stake by value. The
    standard allows transferFrom to use other unspecified mechanisms beyond allowances,
    but all contracts we have examined use only the allowance mechanism.

3.1.2   The Consensys implementation

The code at Listing 3.2 contains the actual Consensys contract implementation, a contract named
EIP20 that extends the EIP20Interface interface. The operations are conformant to the expected
ERC-20 token behavior just described, except for one "special feature" in the transferFrom
function. Next, we remark some features of Solidity used in the code and major details in the
implementation:

- The code begins with the declaration of a number of attributes (lines 2–7), in addition to
  totalSupply inherited from EIP20Interface:

  - MAX_UINT256 is the $2^{256} - 1$ constant, the maximum possible value for a uint256
    expression;

  - balances is a mapping from addresses to owned tokens, i.e., balances[owner] stores the
    amount of tokens owned by owner and is the value returned by a call to balanceOf(
    owner), the function defined in the contract at line 33.

  - allowed is a mapping from addresses to allowances of tokens, in turn expressed as
    another mapping of addresses to allowance values, i.e., allowed[owner][spender] stores
    the allowance of spender in respect to tokens owned by owner and is the value returned
    by a call to allowances(owner,spender), the function defined in the contract at line 42;

  - name, symbol and decimals define some contract properties of informative nature which
    are optional in ERC-20 contracts, respectively defining the contract's name, symbol,
    and decimal scale for tokens.

Listing 3.2: The Consensys contract, a reference implementation of ERC-20.

```solidity
1  contract EIP20 is EIP20Interface {
2    uint256 constant private MAX_UINT256 = 2**256 - 1;
3    mapping (address => uint256) public balances;
4    mapping (address => mapping (address => uint256)) public allowed;
5    string public name;
6    uint8 public decimals;
7    string public symbol;
8    function EIP20(uint256 _initialAmount, string _tokenName,
9                   uint8 _decimalUnits, string _tokenSymbol) public {
10     balances[msg.sender] = _initialAmount;  totalSupply = _initialAmount;
11     name = _tokenName; decimals = _decimalUnits; symbol = _tokenSymbol;
12   }
13   function transfer(address _to, uint256 _value)
14   public returns (bool success) {
15     require(balances[msg.sender] >= _value);
16     balances[msg.sender] -= _value;
17     balances[_to] += _value;
18     emit Transfer(msg.sender, _to, _value);
19     return true;
20   }
21   function transferFrom(address _from, address _to, uint256 _value)
22   public returns (bool success) {
23     uint256 allowance = allowed[_from][msg.sender];
24     require(balances[_from] >= _value && allowance >= _value);
25     balances[_to] += _value;
26     balances[_from] -= _value;
27     if (allowance < MAX_UINT256) {
28         allowed[_from][msg.sender] -= _value;
29     }
30     emit Transfer(_from, _to, _value);
31     return true;
32   }
33   function balanceOf(address _owner) public view returns (uint256 balance) {
34     return balances[_owner];
35   }
36   function approve(address _spender, uint256 _value)
37   public returns (bool success) {
38     allowed[msg.sender][_spender] = _value;
39     emit Approval(msg.sender, _spender, _value);
40     return true;
41   }
42   function allowance(address _owner, address _spender)
43   public view returns (uint256 remaining) {
44     return allowed[_owner][_spender];
45   }
46 }
```

- The contract's constructor consists of a function named after the contract (EIP20) that is executed just once when the contract is deployed to a blockchain. As shown in the code (lines 8–12), the constructor takes in values for all constant attributes including the total supply of tokens. The specification does not dictate how should the supply of tokens be initially allocated, but the usual convention is that that all tokens are allocated to the address of the contract creator, also known as the contract owner, that is given by the caller of the constructor. In the constructor we have in correspondence that balances[msg.sender] as well as totalSupply are initialized to the token supply argument (_initialAmount, at line 10).

- The code of transfer (lines 13–20) illustrate typical ways in which:

  - method preconditions are verified, in this case using the requires statement[3]. – the statement evaluates a Boolean condition and reverts the transaction if the condition does not hold;

  - state attributes are updated, in this case balance;

  - and events are fired using emit, in this case for a Transfer event.

- The code of the contract behaves in line with our previous discussion, except for a "special feature" in transferFrom that is peculiar to the Consensys implementation. Note that the allowance value is only decremented if it is lower than (in practice, it differs from) MAX_UINT256 at line 28. In the code, MAX_UINT256 is used as a "special value" to signal unlimited allowance. The behavior of the contract is deviant in the sense of a "normal" implementation that may allow an improbable but possible allowance of MAX_UINT256 tokens that could be decremented progressively, or even at once in the edge case where the balance of the owner, the token's total supply and the allowance were all equal to MAX_UINT256.

3.1.3   Bug examples

We now illustrate a few real-world bugs in ERC-20 contracts, including deviations to the standard in terms of "calling discipline" (e.g. absent return values, reverts, or events) to bad validation of function pre-conditions leading to the dismissal of valid operations, or, even worst, allowing invalid operations that corrupt a contract's state.

The first set of examples is provided in Listing 3.3. The code shown is taken from the contract of Internet Node Token (INT), and a few bugs are identified in the listing (with BUG):

- The _transfer internal function, called internally by transfer and transferFrom, will revert when balance[_from] == value at line 4. This means that it is not possible to transfer all

---

[3]A equivalent variant is to use if (! precondition) revert (); or, in older versions of Solidity, if (! precondition) throw;.

tokens from an account. The correct pre-condition is balanceOf[_from] >= _value) not balanceOf[_from] > _value (> is used instead of >=).

- The transfer function does not return any value at line 12. It should return true!

- Similarly to the bug in _transfer, we have a boundary check issue in transferFrom at line 17. It is not possible to perform a transfer with an amount that is exactly equal to the allowance at stake. The pre-condition check is _value < allowance[_from][msg.sender], when it should be _value <= allowance[_from][msg.sender] (< is used instead of <=).

- The approve function does not emit any Approval event at line 25, as required!

Listing 3.3: Example bugs in the INT contract.

```
1  function _transfer(address _from, address _to, uint _value) internal {
2    require (_to != 0x0);
3    // BUG: bad pre-condition check
4    require (balanceOf[_from] > _value);
5    require (balanceOf[_to] + _value > balanceOf[_to]); // overflow check
6    balanceOf[_from] -= _value;
7    balanceOf[_to] += _value;
8    Transfer(_from, _to, _value);
9  }
10 function transfer(address _to, uint256 _value) {
11   _transfer(msg.sender, _to, _value);
12   // BUG: no return value
13 }
14 function transferFrom(address _from, address _to, uint256 _value)
15 returns (bool success) {
16   // BUG: bad pre-condition check
17   require (_value < allowance[_from][msg.sender]);
18   allowance[_from][msg.sender] -= _value;
19   _transfer(_from, _to, _value);
20   return true;
21 }
22 function approve(address _spender, uint256 _value)
23 returns (bool success) {
24   allowance[msg.sender][_spender] = _value;
25   // BUG: no Approval event emitted
26   return true;
27 }
```

The second example is taken from the FuturXe contract, shown in Listing 3.4. It includes the bug reported through CVE-2018-12025 [11], that results from an inverted pre-condition check at line 5: we should have allowed[from][msg.sender] < value instead of allowed[from][msg.sender] >= value. Valid transfers are denied, and invalid ones are allowed! For a transferFrom (Alice,Bob,123) call issued by Eve with a 0-token allowance from by Alice, Eve will be able to transfer 123 tokens from Alice's account onto Bob's. The contract's state will subsequently be corrupted in terms of allowances and balances. The code also illustrates a bad pattern for checking pre-conditions, in the sense that it returns false for failed pre-conditions instead of reverting the transaction. The use of require would ensure that any possible changes made to the contract's state are undone, while return false is interpreted as normal control flow that may allow unintended changes to persist.

Listing 3.4: Example bugs in the FuturXe contract.

```solidity
function transferFrom(address from, address to, uint value) returns (bool
    success) {
  if(frozenAccount[msg.sender]) return false;
  if(balances[from] < value) return false;
  // BUG: inverted pre-condition
  if(allowed[from][msg.sender] >= value ) return false;
  if(balances[to] + value < balances[to]) return false;
  balances[from] -= value;
  allowed[from][msg.sender] -= value;
  balances[to] += value;
  Transfer(from, to, value);
  return true;
}
```

3.2 Property-based testing using Brownie

Brownie [3] supports two forms of PBT through Hypothesis, normally called stateless and stateful testing. Stateless testing is employed for tests that are data-driven, have a fixed interaction with the software, and need to model the expected state. In contrast, stateful tests are driven by interactions specified using rule-based state machines, a test case consists of a sequence of invocations of such rules with variable length, and require the current state of the software to be modelled.

We next provide an explanation of the two forms of PBT, in particular the stateful PBT approach we use for ERC-20 contracts, and then explain how test execution works in terms of test lifecycle and the interaction with a test blockchain.

3.2.1 Stateless PBT

Stateless PBT is illustrated by the Python code in Listing 3.5, a simple example adapted from the Brownie documentation. In the code, it is implicitly assumed that contract is an ERC-20 token that is initialized with all tokens associated to accounts[0], where accounts is a container object maintained by Brownie for the accounts created in the test blockchain. The code structure is very similar to what you would expect for a unit test: the transfer function is called for a token contract, and the code then verifies that the source and target accounts are updated correctly.

Listing 3.5: An example of stateless PBT using Brownie.

```
1  from brownie import accounts
2  from brownie.test import given, strategy
3  from hypothesis import settings
4  @given(
5      to=strategy('address', exclude=accounts[0]),
6      value=strategy('uint256', max_value=10000),
7  )
8  @settings(max_examples=100)
9  def test_transfer_amount(contract, to, value):
10     balance = contract.balanceOf(accounts[0])
11     token.transfer(to, value, {'from': accounts[0]})
12     assert contract.balanceOf(accounts[0]) == balance - value
13     assert contract.balanceOf(to) == value
```

But in fact we have a parametric test that may be instantiated with multiple input values that are generated automatically for to and value. These inputs are parameterised by input generation strategies with the @given annotation (at line 4): to is any account except accounts[0], and value will be a random uint256 value with values ranging from 0 to 1000. The @settings annotation (at line 8) indicates that 100 such test examples should be generated at most.

3.2.2 Stateful PBT

Stateless PBT is useful to repeat the same call sequence with various input values, for which Brownie and Hypothesis provide ample support in terms of input generation strategies. Hence, we can think of using it to test ERC-20 contract functions individually, or maybe even a fixed chained sequence. This is not useful however to model an arbitrary sequence of function invocations. In stateful PBT, rule-based state machines define transition rules where each rule exercises the code of a contract, while maintaining a model of its expected state, and assertions verify if the actual state conforms to the expected state. Such rules may be composed in arbitrary sequences with a parameterised value for their maximum length.

Listing 3.6 provides an example of stateful PBT, again adapted from the Brownie documentation. The contract being tested is not an ERC-20 one, instead it provides deposit and withdrawal functions that manipulate ether (rather than tokens) in association to accounts, whose balance can be consulted using deposited. In correspondence to the contract state-changing functions, we have the rule_deposit and rule_withdraw rules that are parameterised by input generation strategies for value and address, in similar manner to stateless PBT. The rule arguments are parameterised by the strategies at lines 5–6. Each rule exercises a contract function that changes the value in deposit for an account, and then asserts that the expected state, modelled by the state-machine variable deposits, matches the actual contract state indicated by the deposited function.

Listing 3.6: An example of stateful PBT using Brownie.

```python
import brownie
from brownie.test import strategy

class StateMachine:
    value = strategy('uint256', max_value="1 ether")
    address = strategy('address')
    def __init__(cls, accounts, Depositer):
        cls.accounts = accounts
        cls.contract = Depositer.deploy({'from': accounts[0]})
    def setup(self):
        self.deposits = {i: 0 for i in self.accounts}
    def rule_deposit(self, address, value):
        self.contract.deposit_for(address, {'from': self.accounts[0], 'value':
            value})
        self.deposits[address] += value
        assert self.deposits[address] == self.contract.deposited(address)
    def rule_withdraw(self, address, value):
        if self.deposits[address] >= value:
            self.contract.withdraw_from(value, {'from': address})
            self.deposits[address] -= value
            assert self.deposits[address] == self.contract.deposited(address)
        else:
            with brownie.reverts():
                self.contract.withdraw_from(value, {'from': address})
def test_stateful(Depositer, accounts, state_machine):
    state_machine(StateMachine, accounts, Depositer,
                settings=settings(max_examples=5000,stateful_step_count=10))
```

Listing 3.7: Possible test output for stateful PBT example.

```
Falsifying example:
state = BrownieStateMachine()
state.rule_deposit(address=<Account '0
    x33A4622B82D4c04a53e170c638B944ce27cffce3'>, value=1)
state.rule_withdraw(address=<Account '0
    x33A4622B82D4c04a53e170c638B944ce27cffce3'>, value=0)
state.teardown()
...
>           assert self.deposits[address] == self.contract.deposited(address)
E           AssertionError: assert 1 == 0
```

The code of **rule_withdraw** illustrates an additional common testing pattern in Brownie. The rule deals with two cases, depending on whether the pre-condition self.deposits[address] >= value for **withdraw** at line 17 holds or not for particular values of **address** and **value**. When the condition is met, the code expects a normal interaction with the blockchain and a correct change in the contract state (lines 18–20). Otherwise, the code expects a transaction revert (lines 22–23) as expressed by the block of code starting with with brownie. reverts (). When a revert is expected and the contract fails to do so, an assertion error is fired.

During test execution, the rules in the example state machine can be called in sequence, possibly more than once per rule and in any order per each test example. If there is a bug in the contract, a falsifying example will correspond to one such sequence. For instance one could obtain the Brownie output shown in Listing 3.7, where the falsifying example details a sequence composed by a deposit and a withdraw and the corresponding values used for value and address in each rule invocation. As in stateless PBT, the maximum number of test examples is configurable, but stateful PBT is also parameterised by a "stateful step count" parameter indicating the maximum number of rules to execute for a single test example. The two settings are illustrated in the code at line 26.

3.2.3 Test execution

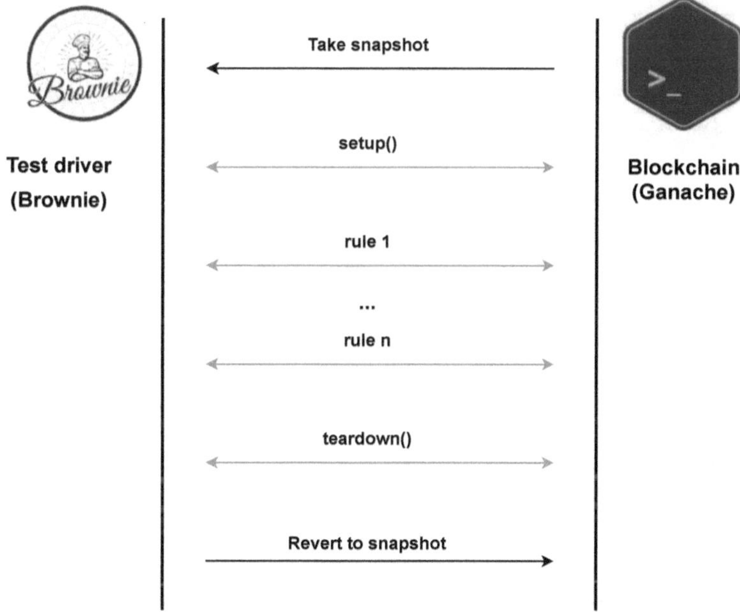

Figure 3.1: Brownie stateful testing flow

Brownie executes a test case according to fixtures that can be defined for the test lifecycle, coupled with test isolation mechanisms necessary for reproducible testing. Brownie employs the Ganache [24] blockchain during tests, a popular choice for Ethereum development environments (e.g., Truffle [55] also employs Ganache). Ganache is an in-memory blockchain that can be setup on-the-fly with Ethereum accounts and contracts, and supports a snapshot mechanism that allows Brownie to reset the blockchain back to a desired state.

The overall process of test execution is illustrated in Figure 3.1, and can be described as follows:

- **One-time initialization**

  - Brownie starts, booting Ganache with an initial configuration for accounts and ether balances.

  - The test class constructor is constructor. This will be the state machine constructor (__init__) for stateful PBT. The constructor is responsible for deploying the contracts to test, for instance as in line 9 of Listing 3.6, along with other one-time setup actions for the blockchain and the test logic.

  - Brownie takes a snapshot of the Ganache blockchain. This snapshot state can then be re-instated after each test, as discussed below.

- **Test execution** (per each example generated internally by Hypothesis)

  - The setup() method, if defined, is called to initialise the test logic. In stateful PBT, this step can be used to (re-)initialise model variables that are changed during rule execution.

  - The test example is executed. For stateful PBT, this will correspond to the execution of a sequence of rule methods in the state machine. For stateless PBT, a single test method is executed.

  - The teardown() method, if defined, is called to tear down the test logic, regardless of whether the execution test example failed or not.

  - Brownie reverts the Ganache blockchain to the snapshot set during one-time initialisation.

3.3   PBT state machine for ERC-20

3.3.1   Overview

Our PBT framework for ERC-20 contracts is based on the definition of a Brownie state machine. The skeleton is provided in Listing 3.8. As shown, the StateMachine class defines the typical lifecycle methods discussed earlier (__init__, setup, and teardown), 5 rule methods, and some auxiliary methods for test assertions. There is one rule per each of the ERC-20 state-changing functions – rule_approve, rule_transfer , rule_transferFrom – plus two other rules that exercise special edge cases – rule_transferAll and rule_approveAndTransferAll. We illustrate the major aspects of these definitions with a few code samples in this section (the full source code is listed in Appendix A).

Listing 3.8: ERC-20 state machine – overview of methods.

```
class StateMachine:
  # Input generation strategies
  ...
  # Test lifecycle methods
  def __init__(self, accounts, contract, totalSupply, DEBUG=None):
    ...
  def setup(self)
    ...
  def teardown(self)
    ...
  # Rules
  def rule_transfer(self, st_sender, st_receiver, st_amount):
    ...
  def rule_transferFrom(self, st_spender, st_owner, st_receiver, st_amount):
    ...
  def rule_approve(self, st_owner, st_spender, st_amount):
    ...
  def rule_transferAll(self, st_sender, st_receiver):
    ...
  def rule_approveAndTransferAll(self, st_owner, st_spender, st_receiver):
    ...
  # Auxiliary assertion methods
  def verify_TotalSupply(self):
    ...
  def verifyAllBalances(self):
    ...
  ... other assertion methods ...
```

3.3.2 Base logic

The StateMachine code for input generation strategies and test lifecycle methods is shown in Listing 3.9. The main aspects are as follows:

- Input generation strategies (lines 2–6) are defined for token amounts and addresses for the possible various roles in a ERC-20 function (sender or caller address, receiver, and owner).

- The state machine constructor (lines 7–11) takes as arguments the list of blockchain accounts (accounts), a contract deployed in the blockchain (contract), the token's total supply (totalSupply), and a debug output flag (DEBUG). These properties are stored in corresponding attributes for a StateMachine instance. The constructor should be called

Listing 3.9: ERC-20 state machine – input generation strategies, lifecycle methods and state modelling.

```python
1  class StateMachine:
2    st_amount = strategy("uint256")
3    st_owner = strategy("address")
4    st_spender = strategy("address")
5    st_sender = strategy("address")
6    st_receiver = strategy("address")
7    def __init__(self, accounts, contract, totalSupply, DEBUG=None):
8      self.accounts = accounts
9      self.contract = contract
10     self.totalSupply = totalSupply
11     self.DEBUG = DEBUG
12     ...
13   def setup(self):
14     if self.DEBUG:
15       print("setup()")
16     self.allowances = dict() :
17     self.balances = {i: 0 for i in self.accounts} :
18     self.balances[self.accounts[0]] = self.totalSupply
19     self.value_failure = False
20   def teardown(self):
21     if self.DEBUG:
22       print("teardown()")
23     if not self.value_failure:
24       self.verifyTotalSupply()
25       self.verifyAllBalances()
26       self.verifyAllAllowances()
27 ...
```

by constructors of **StateMachine** sub-classes, that should take care first of creating and deploying the contract to be tested, as we detail further on in this section.

- The **setup()** method (lines 13–19) initialises instance variables to model **allowances** and **balances**, corresponding to the initial state of a contract. In relation, the following invariants must always hold: **balances**[a] should always equal **contract**.**balanceOf**(a) for every address a, and, similarly, **allowances**[a][b] should always equals **contract**.**allowances**(a,b) for every pair of addresses a and b.

- The **teardown()** method (lines 20–26) performs a complete verification of the entire state of the contract, as long as a previous assertion on values has not failed. As we will see, rules perform assertions as they execute but limited to the accounts they have operated on, hence this final verification ensures that the model invariants hold for all accounts.

Listing 3.10: State machine rules.

```python
def rule_transfer(self, st_sender, st_receiver, st_amount):
  if self.DEBUG:
    print("transfer({}, {}, {})".format(st_sender, st_receiver, st_amount))
  if st_amount <= self.balances[st_sender]:
    with normal():
      tx = self.contract.transfer(st_receiver, st_amount,
                                  {"from": st_sender})
      self.verifyTransfer(st_sender, st_receiver, st_amount)
      self.verifyEvent(tx,"Transfer",{"from": st_sender, "to": st_receiver,
                                      "value": st_amount})
      self.verifyReturnValue(tx, True)
  else:
    with brownie.reverts():
      self.contract.transfer(st_receiver, st_amount, {"from": st_sender})
def rule_transferFrom(self, st_spender, st_owner, st_receiver, st_amount):
  ...
def rule_approve(self, st_owner, st_spender, st_amount):
  if self.DEBUG:
    print("approve({}, {}, {})".format(st_owner, st_spender, st_amount))
  with normal():
    tx = self.contract.approve(st_spender, st_amount, {"from": st_owner})
    self.verifyAllowance(st_owner, st_spender, st_amount)
    self.verifyEvent(tx,"Approval",{"owner": st_owner, "spender": st_spender
      , "value": st_amount})
    self.verifyReturnValue(tx, True)
def rule_transferAll(self, st_sender, st_receiver):
  self.rule_transfer(st_sender, st_receiver, self.balances[st_sender])
def rule_approveAndTransferAll(self, st_owner, st_spender, st_receiver):
  amount = self.balances[st_owner]
  self.rule_approve(st_owner, st_spender, amount)
  self.rule_transferFrom(st_spender, st_owner, st_receiver, amount)
```

3.3.3 Rules

Listing 3.10 provides the code for the rules in the state machine, except for rule_transferFrom
for space reasons.

Beginning with rule_transfer (lines 1–14), we see that it accounts for two cases, depending on
whether the st_amount <= self.balances[st_sender] (line 4) holds or not on entry. The condition
models whether st_sender has enough balance to transfer st_amount tokens to st_receiver, the
pre-condition imposed for transfer in the ERC-20 specification. If the condition holds (lines 5–11),
then the contract function is called, and, subsequently, it is verified if the transfer took place
between accounts st_sender and st_receiver (through the call to verifyTransfer), that a Transfer
event has been emitted (through verifyEvent), and, finally, that transfer returned true (through
verifyReturnValue). If the transfer pre-condition does not hold, then a transaction revert is

expected (lines 13–14). The code for rule_transferFrom, omitted due to space reasons, overall follows the pattern of rule_transfer , even if slightly more complex in regard to pre-conditions and verifications to perform.

The code for rule_approve (lines 17–24) is simpler than that of rule_transfer , given that approve has no pre-conditions. The rule proceeds by calling the contract's approve method, and then verifying that the corresponding allowance update took place (in the call to verifyAllowance), an Approval event has been fired, and finally that the function returned true.

The final two rules, rule_transferAll and rule_approveAndTransferAll, can be seen as "derived" given that they work by invoking the other three rules. The purpose is to exercise the edge case of transferring all tokens owned by an account through transfer or transferFrom more easily during testing. The rule_transferAll rule (lines 25-26) tests the transfer of all tokens belonging to an account onto another account. As shown, the code consists of an invocation of rule_transfer such that the amount of tokens equals all of the tokens owned by st_sender. As for rule_approveAndTransferAll (lines 27-30), it tests the transfer of all tokens again but through transferFrom. The code contains a call to rule_approve, to set the required allowance first, followed by a call to rule_transferFrom.

Listing 3.11: Auxiliary methods used for verification.

```python
def verifyBalance(self, addr):
    self.verifyValue("balanceOf({})".format(addr),
                     self.balances[addr],
                     self.contract.balanceOf(addr))
def verifyTransfer(self, src, dst, amount):
    self.balances[src] -= amount
    self.balances[dst] += amount
    self.verifyBalance(src)
    self.verifyBalance(dst)
def verifyAllowance(self, owner, spender, delta=None):
    if delta != None:
        if delta >= 0:
            self.allowances[(owner,spender)] = delta
        elif delta < 0:
            self.allowances[(owner,spender)] += delta
    self.verifyValue("allowance({},{})".format(owner, spender),
                     self.allowances[(owner, spender)],
                     self.contract.allowance(owner, spender))
def verifyValue(self, msg, expected, actual):
    if expected != actual:
        self.value_failure = True
        raise AssertionError(
            "{} : expected value {}, actual value was {}".format(msg,
                                                                 expected,
                                                                 actual))
```

3.3.4   Verification methods

Finally, we describe the utility methods in the state machine that are used for updates to the model state, and also perform assertions that compare the actual state of contract with the model state.

The most relevant methods involve the balances and allowances model variables, and are shown in Listing 3.11. Calls to verifyTransfer (lines 5–9) and verifyAllowance (lines 10–18) lead to the update of model variables balances and allowances, respectively. These methods subsequently verify that the contract's state, as reported by the balanceOf or allowance contract functions, conforms to the contents of the model variables. The verifyValue method (lines 19–25), used as fallback by other verification methods, throws an AssertionError exception in case a value reported by the contract does not match the expected one.

3.3.5   State machine extensions

In addition to the base ERC-20 functionality, many contracts usually provide other functionalities. For instance, it is quite common to find contracts support frozen accounts, transfer of ownership, or contract pausing. Our attention focused on three functionalities that involve manipulation of tokens: minting, burning, and buy/sell operations in which tokens can be obtained from or exchanged to ether. Token minting corresponds to the creation of tokens, increasing the total supply of tokens and associating the newly minted tokens to some address. Token burning is the reverse operation: tokens can be erased from an account and their total supply decreases. Token sale works in terms of operations that allow an account to buy tokens using ether, or obtain ether by selling tokens.

We devised extensions of StateMachine that account for these operations, extending the base state machine for ERC-20 operations. The implementation of state machine variants could not be guided by a strict specification, however, but informally by the analysis of a set of real-world contracts (covered in the evaluation of Chapter 5), and the perceived coherence of operations with the spirit of ERC-20. We should note that most contracts do not implement all 3 kinds of functionality, in fact only one of the contracts analysed does so, and that token minting and burning are not always implemented both by contracts. This leads to the following model assumptions, incorporated in three subclasses of StateMachine:

- **Token minting** – a call to mintToken(receiver, amount), when issued by the contract's owner, increases both the token's total supply and the balance of receiver by amount, as long as the current total supply plus amount do not exceed $2^{256} - 1$, the maximum possible amount of tokens. The void operation of minting 0 tokens should be allowed, in line with ERC-20 that similarly allows 0-token transfers. The code for the state machine extension, MintingStateMachine, is provided in Listing 3.12. Note that the parent class is StateMachine, hence all rules for the standard ERC-20 functions are inherited. In addition,

Listing 3.12: State machine extension for token minting.

```python
class MintingStateMachine(StateMachine):
    def __init__(self, accounts, contract, totalSupply, DEBUG=None):
        StateMachine.__init__(self, accounts, contract, totalSupply, DEBUG)
    def rule_mint(self, st_receiver, st_amount):
        if self.DEBUG:
            print("mint({}, {})".format(st_receiver, st_amount))
        if st_amount + self.totalSupply <= 2 ** 256 - 1:
            with normal():
                self.contract.mintToken(st_receiver, st_amount,
                                        {"from": self.accounts[0]})
                self.totalSupply += st_amount
                self.balances[st_receiver] += st_amount
                self.verifyBalance(st_receiver)
                self.verifyTotalSupply()
        else:
            with (brownie.reverts()):
                self.contract.mintToken(st_receiver, st_amount,
                                        {"from": self.accounts[0]})
```

Listing 3.13: State machine extension for token burning.

```python
class BurningStateMachine(StateMachine):
    def __init__(self, accounts, contract, totalSupply, DEBUG=None):
        StateMachine.__init__(self, accounts, contract, totalSupply, DEBUG)
    def rule_burn(self, st_sender, st_amount):
        if self.DEBUG:
            print("burn({}, {})".format(st_sender, st_amount))
        if st_amount >= 0 and self.balances[st_sender] >= st_amount:
            with normal():
                tx = self.contract.burn(st_amount, {"from": st_sender})
                self.totalSupply -= st_amount
                self.balances[st_sender] -= st_amount
                self.verifyBalance(st_sender)
                self.verifyTotalSupply()
                self.verifyEvent(tx, "Burn", {"from": st_sender, "value": st_amount})
            else:
                with (brownie.reverts()):
                    self.contract.burn(st_amount, {"from": st_sender})
    def rule_burn_all(self, st_sender):
        self.rule_burn(st_sender, self.balances[st_sender])
```

Listing 3.14: BuySellStateMachine

```python
class BuySellStateMachine(StateMachine):
  INITIAL_BUY_PRICE = 1
  INITIAL_SELL_PRICE = 1
  def __init__(self, accounts, contract, totalSupply, DEBUG=None):
    StateMachine.__init__(self, accounts, contract, totalSupply, DEBUG)
  ...
  def setup(self):
    ...
    self.ethBalances = {i: i.balance() for i in self.accounts}
    self.ethBalances[self.contract] = self.contract.balance()
  def rule_setPrices(self, st_amount):
    ...
  def rule_sell(self, st_sender, st_amount):
    if self.DEBUG:
      print("sell({}, {})".format(st_sender, st_amount))
    ether = st_amount * self.sellPrice
    if self.balances[st_sender] >= st_amount and self.ethBalances[self.
        contract] >= ether:
      with normal():
        tx = self.contract.sell(st_amount, {"from": st_sender})
        self.verifySale(st_sender, self.contract, st_amount, ether, tx)
    else:
      with (brownie.reverts()):
        self.contract.sell(st_amount, {"from": st_sender})
  def rule_buy(self, st_sender, st_amount):
    if self.DEBUG:
      print("buy({}, {})".format(st_sender, st_amount))
    if self.buyPrice > 0 and self.ethBalances[st_sender] >= st_amount\
        and self.balances[self.contract] >= st_amount // self.buyPrice):
      with normal():
        tx = self.contract.buy({"from": st_sender, "value": st_amount})
        self.verifySale(self.contract, st_sender,
                        st_amount // self.buyPrice, st_amount, tx)
    elif self.ethBalances[st_sender] >= st_amount:
      with (brownie.reverts()):
        self.contract.buy({"from": st_sender, "value": st_amount})
  def rule_sellAll(self, st_sender):
    self.rule_sell(st_sender, self.balances[st_sender])
  def verifyEthBalance(self, addr):
    self.verifyValue(
        "ethBalance({})".format(addr),
        self.ethBalances[addr],
        addr.balance())
  def verifySale(self, a, b, tokens, ether, tx):
    self.balances[a] -= tokens
    self.balances[b] += tokens
    self.ethBalances[a] += ether
    self.ethBalances[b] -= ether
    self.verifyBalance(a)
    self.verifyBalance(b)
    self.verifyEthBalance(a)
    self.verifyEthBalance(b)
```

the rule_mint rule is defined.

- **Token burning** – a call to burn(amount) decrements both the total supply of tokens and
  the balance of the caller's address (msg.sender) buy amount, as long as the balance of the
  caller has at least amount tokens. Similarly to token minting, burning 0 tokens should be
  allowed. The code for the state machine extension, BurningStateMachine, is provided in
  Listing 3.13. It defines the base rule_burn rule and the rule_burnAll "derived" rule that
  explicitly tests the burning of all tokens in an account.

- **Token sale** – we found that contracts implement three functions, as follows:

  - A call to setPrices (buyPrice, sellPrice) set the token's buy and sell prices in terms
    of ether. From the contracts analysed, it becomes unclear what should happen when
    any of these prices is set to 0 though, as the sell and buy functions detailed below do
    not perform any check for a 0 value, or revert in that case as it would seem reasonable.

  - A call to sell (amount) allows the caller to sell amount tokens from its balance, and
    obtain in return amount ∗ sellPrice units of ether. The tokens are transferred in that
    case to the address of the contract itself. The function should revert if the sell price is
    0, or the caller has an insufficient amount of tokens, or the contract has no ether to
    pay the caller.

  - A call to buy() with an associated ether value amount passed on as implicit argument
    should increase the caller's token balance by amount / buyPrice. The function should
    revert if the buy price is 0, or the balance associated to the contract's address has
    insufficient tokens.

The code for the state machine extension, BuySellStateMachine, is partially provided in
Listing 3.13. There is a rule per each of the functions described above plus the rule_sellAll
"derived rule" that explicitly tests the sale of all tokens. An interesting aspect of this state
machine extension is the need to model ether balances that are associated with Ethereum
accounts in addition to token balances maintained by an ERC-20 contract. This is done
through model variable ethBalances, initialized in setup (lines 9–10) and updated during
verification of buy/sell operations in verifySale (lines 46–47).

3.4   PBT execution for ERC-20 contracts

3.4.1   Test instantiation and execution

A test script is instantiated for a contract in our framework as shown in Listing 3.15, using the
Consensys token test script as an example.

The code in the example starts with the necessary imports (lines 1–2), one for the pytest
standard Python testing library, and another one for the StateMachine definition from our
erc20_pbt module. The identifier of the ERC-20 contract to test is then specified to be EIP20

Listing 3.15: Example definition of a test script.

```
1  import pytest
2  from erc20_pbt import StateMachine
3
4  @pytest.fixture()
5  def contract2test(EIP20):
6      yield EIP20
7
8  class Consensys(StateMachine):
9      def __init__(self, accounts, contract2test):
10          totalSupply = 1000
11          contract = contract2test.deploy(
12              totalSupply, "Consensys", 10, "XYZ", {"from": accounts[0]}
13          )
14          StateMachine.__init__(self, accounts, contract, totalSupply)
15
16  def test_stateful(contract2test, accounts, state_machine):
17      state_machine(Consensys, accounts, contract2test)
```

Listing 3.16: Usage message for the pbt script.

```
Usage:
  pbt [options] test1 ... testn
Options:
  -c <arg> : set stateful step count
  -n <arg> : set maximum examples
  -s <arg> : set seed for tests
  -C       : measure coverage
  -D       : enable debug output
  -E       : enable verification of events
  -R       : enable verification of return values
  -S       : enable shrinking
```

(lines 4–6), and test class Consensys is defined (lines 8–14) as an extension of the StateMachine class. Alternatively, one of the StateMachine sub-classes for token minting, burning or sale could be specified instead as a parent class. The Consensys class constructor takes care of deploying the contract in the blockchain, specifying a certain total supply tokens along with other parameters that are in turn required for the invocation of the contract constructor, and then feeds the contract already in deployed form to the StateMachine parent class constructor. The script ends with the definition of the test_stateful method (lines 16–17) that will be invoked by the brownie or pytest programs to initiate testing.

To facilitate the invocation of PBT test with several configurable options, we developed a simple shell script called pbt as a wrapper for the invocation of pytest. The pbt usage message is given in Listing 3.16. The invocation options relate to basic PBT parameters (stateful step count, number of examples, test seed, optional shrinking of falsifying examples), and the enabling of operation modes (coverage monitoring and debug output) or optional verification features (events and return values).

3.4.2 Example test executions

We illustrate the execution of the pbt tool with a few fragments of reports produced by it for some bugs in the INT and FuturXe contracts discussed earlier in Section 3.1.3.

Listing 3.17 shows 3 falsifying examples for the INT contract. The first one (lines 1–17) reports an unexpected revert during the execution of rule_approveAndTransferAll, and the corresponding cause of revert. As highlighted earlier in this chapter, the INT contract reverts unexpectedly in transferFrom when the amount being transferred equals the allowance of the spender, due to a bug in a pre-condition check. The failed pre-condition at stake is identified in the PBT report (line 16). The other two falsifying examples relate to the absence of a return value in the execution of transfer through rule_transfer (lines 18–26) and the mission emission of an Approval event in the execution of approve through rule_approveAndTransferAll (lines 27–39).

Listing 3.17 shows 2 falsifying examples for the FuturXe contract, both related to the negated pre-condition bug that does not allow valid transfers through transferFrom, and on the other hand, allows an invalid transfer when no allowance is set in the same function. The first one (lines 1–12) illustrates that the allowance value is not updated for a valid transfer. This happens because the buggy pre-condition check causes transferFrom to return without any updates to the allowance at stake (or to the balances of the owner and receiver account; the first failed assertion halts the test and allowances are verified first, so the invalid balances are not eventually reported). The second one (lines 14–23) relates to the inverse case: no allowance is set but the transfer does take place and updates the allowance and balances. The output signals that a revert was expected.

Listing 3.17: Falsifying examples for INT contract.

```
1  Falsifying example:
2  state = BrownieStateMachine()
3  state.rule_approveAndTransferAll(st_owner=<Account '0
       x33A4622B82D4c04a53e170c638B944ce27cffce3'>, st_receiver=<Account '0
       x33A4622B82D4c04a53e170c638B944ce27cffce3'>, st_spender=<Account '0
       x33A4622B82D4c04a53e170c638B944ce27cffce3'>)
4  state.teardown()
5
6  Traceback (most recent call last):
7    File "/home/edrdo/brownie_test/erc20_pbt.py", line 115, in
         rule_approveAndTransferAll
8      self.rule_transferFrom(st_spender, st_owner, st_receiver, amount)
9    ...
10   File "/home/edrdo/brownie_test/erc20_pbt.py", line 76, in rule_transferFrom
11     tx = self.contract.transferFrom(
12 brownie.exceptions.VirtualMachineError: revert
13 Trace step −1, program counter 1476:
14   File "contracts/INT.sol", line 67, in token.transferFrom:
15 ...
16         require (_value < allowance[_from][msg.sender]);
17 ...
18 Falsifying example:
19 state = BrownieStateMachine()
20 state.rule_transfer(st_amount=168, st_receiver=<Account '0
       x66aB6D9362d4F35596279692F0251Db635165871'>, st_sender=<Account '0
       x66aB6D9362d4F35596279692F0251Db635165871'>)
21 state.teardown()
22
23 Traceback (most recent call last):
24   File "/home/edrdo/brownie_test/erc20_pbt.py", line 56, in rule_transfer
25     self.verifyReturnValue(tx, True)
26 AssertionError: return value : expected value True, actual value was None
27 ...
28 Falsifying example:
29 state = BrownieStateMachine()
30 state.rule_approveAndTransferAll(st_owner=<Account '0
       x66aB6D9362d4F35596279692F0251Db635165871'>, st_receiver=<Account '0
       x66aB6D9362d4F35596279692F0251Db635165871'>, st_spender=<Account '0
       x0063046686E46Dc6F15918b61AE2B121458534a5'>)
31 state.teardown()
32
33 Traceback (most recent call last):
34   File "/home/edrdo/brownie_test/erc20_pbt.py", line 114, in
         rule_approveAndTransferAll
35     self.rule_approve(st_owner, st_spender, amount)
36 ...
37   File "/home/edrdo/brownie_test/erc20_pbt.py", line 102, in rule_approve
38     self.verifyEvent(
39 AssertionError: Approval: event was not fired
```

Listing 3.18: Falsifying examples for FuturXe contract.

```
1  Falsifying example:
2  state = BrownieStateMachine()
3  state.rule_approveAndTransferAll(st_owner=<Account '0
       x66aB6D9362d4F35596279692F0251Db635165871'>, st_receiver=<Account '0
       x66aB6D9362d4F35596279692F0251Db635165871'>, st_spender=<Account '0
       x0063046686E46Dc6F15918b61AE2B121458534a5'>)
4  state.teardown()
5
6  Traceback (most recent call last):
7    File "/home/edrdo/brownie_test/erc20_pbt.py", line 115, in
          rule_approveAndTransferAll
8    ...
9    File "/home/edrdo/brownie_test/erc20_pbt.py", line 81, in rule_transferFrom
10     self.verifyAllowance(st_owner, st_spender, −st_amount)
11 AssertionError: allowance(0x66aB6D9362d4F35596279692F0251Db635165871,0
       x0063046686E46Dc6F15918b61AE2B121458534a5) : expected value 0, actual
       value was 1000
12
13 ...
14 Falsifying example:
15 state = BrownieStateMachine()
16 state.rule_transferFrom(st_amount=1, st_owner=<Account '0
       x66aB6D9362d4F35596279692F0251Db635165871'>, st_receiver=<Account '0
       x0063046686E46Dc6F15918b61AE2B121458534a5'>, st_spender=<Account '0
       x66aB6D9362d4F35596279692F0251Db635165871'>)
17 state.teardown()
18
19 Traceback (most recent call last):
20 ...
21   File "/home/edrdo/brownie_test/erc20_pbt.py", line 90, in rule_transferFrom
22 ...
23 AssertionError: Transaction did not revert
```

Chapter 4

Implementation details

This Chapter covers complementary details for the PBT implementation and the organisation of unit testing suites we use in the evaluation (provided later in Chapter 5). We first describe the project structure for PBT, along with implementation details concerning test execution, and a few necessary patches applied to Brownie and Hypothesis (Section 4.1). This is followed by a description of the Truffle framework we used for the Consensys and OpenZeppelin unit testing suites (4.2).

4.1 PBT framework

4.1.1 Project organisation

Figure 4.1 depicts the directory structure that concerns the definition of the PBT framework and its use for testing ERC-20 contracts. On the root level of the project we find two files: erc20_pbt.py, containing the PBT implementation, and pbt, the bash script used to execute tests. ERC-20 contracts are placed in sub-directories (e.g., BNB), each of which is a Brownie project containing: contracts/, where the source code files for contracts can be found; tests/, the location for test scripts; and reports/, for storage of coverage reports and logs from test executions.

4.1.2 Test class hierarchy

The organisation of test classes is illustrated in Figure 4.2. As shown, the infrastructure comprises the base support from Hypothesis and Brownie, the ERC-20 state machines of our PBT framework, and, finally, the concrete test instantiations for contracts. The Hypothesis framework for PBT defines the base RuleBasedStateMachine class. Brownie in turn refines this base functionality into a BrownieStateMachine class, that abstracts aspects such as blockchain setup, snapshots, and state rollbacks required by test isolation. Brownie defines its custom form of rule-based state

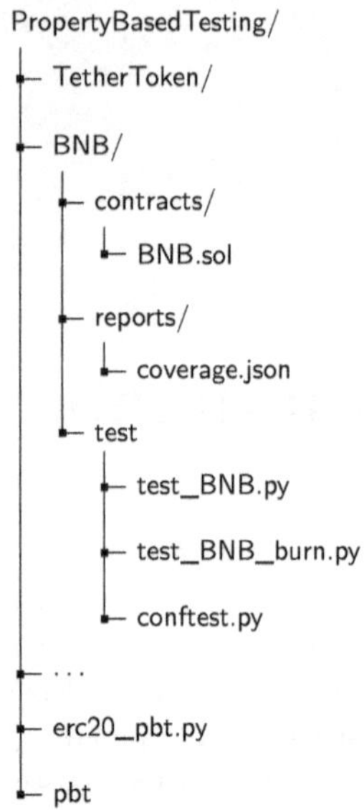

Figure 4.1: Brownie project environment

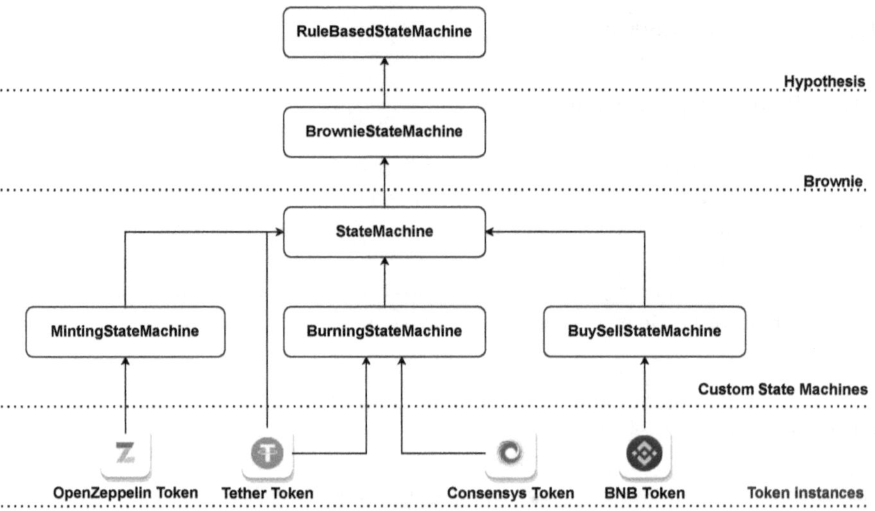

Figure 4.2: Test Class Hierarchy

machines, of which the ERC-20 StateMachine is an example. Our PBT framework additionally defines MintingStateMachine, BurningStateMachine, and BuySellStateMachine as extensions of

StateMachine. All the state machine classes in the PBT framework can then be subclassed for contracts of interest through test scripts.

4.1.3 Hypothesis settings and profiles

Listing 4.1: Register Hypothesis profiles

```python
def register_hypothesis_profiles():
    import hypothesis
    from hypothesis import settings, Verbosity, Phase
    stateful_step_count = int(os.getenv("PBT_STATEFUL_STEP_COUNT", 10))
    max_examples = int(os.getenv("PBT_MAX_EXAMPLES", 100))
    derandomize = True
    seed = int(os.getenv("PBT_SEED", 0))
    if seed != 0:
        patch_hypothesis_for_seed_handling(seed)
        derandomize = False
    patch_brownie_for_assertion_detection()
    settings.register_profile(
        "generate",
        stateful_step_count=stateful_step_count,
        max_examples=max_examples,
        phases=[Phase.generate],
        report_multiple_bugs=True,
        derandomize=derandomize,
        print_blob=True)
    settings.register_profile(
        "shrinking",
        stateful_step_count=stateful_step_count,
        max_examples=max_examples,
        phases=[Phase.generate, Phase.shrink],
        report_multiple_bugs=True,
        derandomize=derandomize,
        print_blob=True)
```

Hypothesis has a set of parameters (Hypothesis. settings) that control the PBT process:

- derandomize=True|False controls if tests run deterministically. If True, the examples will be generated deterministically thus allowing for reproducible test executions, using a fixed seed that is generated internally from a signature of the test class. If False, tests will run deterministically only if a seed is explicitly set for the state machine, (a process we described later in this chapter), otherwise a seed is picked at random.

- max_examples=n defines how many examples will be generated.

- stateful_step_count=n defines the maximum number of rules that can be invoked per each test example.

- **phases** controls which testing phases should be run. The phases of interest in this work are the **generate** phase, which simply tells Hypothesis to only generate examples for test, and the **shrink** phase, which upon a failure tries to simplify each falsifying example to a minimal simplified version that can replicate the exact same failure.

- **report_multiple_bugs=True|False** tells Hypothesis whether to stop execution after the first falsifying example (bug) is found, or if it may continue and report multiple ones.

- **print_blob=True|False** will if enabled instruct Hypothesis to print code for failing examples that can be used to reproduce those examples later.

A combination of the settings above forms what is called a profile in Hypothesis. For our PBT framework, we defined a `register_hypothesis_profiles` () method shown in Listing 4.1. This method is activated through the `conf_test.py` file in each contract's **test** **directory** that is automatically executed by **pytest**. It registers two profiles that can be instantiated for PBT test execution, **"generate"** and **"shrinking"**, and performs other parameterization actions with inputs from environment variables set by the **pbt** script. The **"generate"** profile (lines 12–19) takes configurable values for the maximum number of examples and stateful step count, and tells Hypothesis to go through the example generation phase, report multiple bugs, and print blobs. The **"shrinking"** profile (lines 20–27) has the same definitions, but additionally instructs Hypothesis to shrink falsifying examples.

4.1.4   Adjustments to Hypothesis and Brownie

During development, we found a few limitations in the implementations of Brownie and Hypothesis that were dealt with using cirurgical patches/instrumentations to their code. The adjustments were necessary to the multiple bug reporting and seed handling mechanisms of Hypothesis, and the revert-handling mechanism of Brownie. We provide a summary of these adjustments next.

4.1.4.1   Multiple bug reporting

Listing 4.2: Hypothesis example generation capped

```
1 == Original code
2 < """ We cap 'calls after first bug' so errors are reported reasonably
3 < soon even for tests that are allowed to run for a very long time,
4 < or sooner if the latest half of our test effort has been fruitless."""
5 < return self.call_count < MIN_TEST_CALLS or self.call_count < min(
6     self.first_bug_found_at + 1000, self.last_bug_found_at * 2)
7 == Modification
8 > return self.valid_examples <= self.settings.max_examples \
9         and self.call_count <= 2 * self.settings.max_examples
```

Hypothesis reports multiple bugs, but does not honor the max_examples value after it finds the first bug. We were expecting example generation to continue up until max_examples regardless of the bug count, as in a fuzz testing approach. However, once Hypothesis finds the first bug, the search for more bugs is limited to a fixed number of extra examples. This lead us to apply the patch in the engine.py source file of Hypothesis, shown in Listing 4.2 lines 8-9.

4.1.4.2 Hypothesis seed handling

Listing 4.3: Instrumentation for seed injection.

```
1 def patch_hypothesis_for_seed_handling(seed):
2   import hypothesis
3   h_run_state_machine = hypothesis.stateful.run_state_machine_as_test
4   def run_state_machine(state_machine_factory, settings=None):
5     state_machine_factory._hypothesis_internal_use_seed = seed
6     h_run_state_machine(state_machine_factory, settings)
7   hypothesis.stateful.run_state_machine_as_test = run_state_machine
```

By default, tests will run with derandomize=True and Hypothesis will derive a fixed seed to run the tests with. For our evaluation purposes it was convenient to configure the seed to use when desired. We found out that it is possible to a set a user-defined seed, but only if the state machine at stake conveys the seed through a field called[1] _hypothesis_internal_use_seed . The problem was that Brownie performs some complex adjustments to the state machine representation on-the-fly, which discards the use of that setting within our StateMachine class. The code shown at Listing 4.3 replaces the run_state_machine_as_test method in Hypothesis such that it sets the desired seed value before executing tests.

4.1.4.3 Brownie instrumentation for proper revert detection

When dealing with returned exceptions, Brownie does not differentiate between a revert and another kind of EVM exception, e.g., assertion errors or other virtual machine errors (out of gas errors, divisions by zero, etc). The brownie.reverts() handler will complete successfully when any kind of EVM exception is thrown. More conveniently, it should only succeed if the error corresponds to a revert. We applied the instrumentation shown in Listing 4.4 so that reverts are properly detected, so that Brownie's revert context manager only considers a proper revert to be one where the EVM exception instance has the revert_type field set to "revert" (line 8).

Listing 4.4: Instrumentation for proper revert detection.

```python
1  def patch_brownie_for_assertion_detection():
2      from brownie.test.managers.runner import RevertContextManager
3      from brownie.exceptions import VirtualMachineError
4      f = RevertContextManager.__exit__
5      def alt_exit(self, exc_type, exc_value, traceback):
6          if exc_type is VirtualMachineError:
7              exc_value.__traceback__.tb_next = None
8              if exc_value.revert_type != "revert":
9                  return False
10         return f(self, exc_type, exc_value, traceback)
11     RevertContextManager.__exit__ = alt_exit
```

4.2 Unit testing

4.2.1 Truffle project environment

Truffle is the framework used by the Consensys and OpenZeppelin testing suites and the one we use in our evaluation chapter. Like Brownie, Truffle is a development environment for contracts, i.e., it integrates compilation, testing, and deployment of smart contracts. Tests are writen using Javascript using the Mocha test runner [34] and the Chai assertion library [5].

We assembled the unit testing suites with the organisation shown in Figure 4.3. The root directory contains a parent Truffle project and sub-directories named after contracts contain child projects. The root and child projects may each contain truffle−config .js configuration file and the following subdirectories: contracts/ with the source code of contracts; migrations/ with scripts that are responsible for staging deployment tasks[2]; and test/ with the Javascript source code for unit tests.

The Truffle configuration file is used to set the Solidity compiler version and to explicitly tell Truffle that the current project will make use of solidity−coverage plugin. An example configuration is provided in Listing 4.5.

To support additional plugins like solidity−coverage, chai and mocha at specific versions, this setup makes use of the npm [39] package manager to provision the Truffle project with the required dependencies. The exact packages and their correspondent versions can be seen in Listing 4.6.

---

[2]Migration scripts are a way to tell Truffle about the contracts we want to interact with.

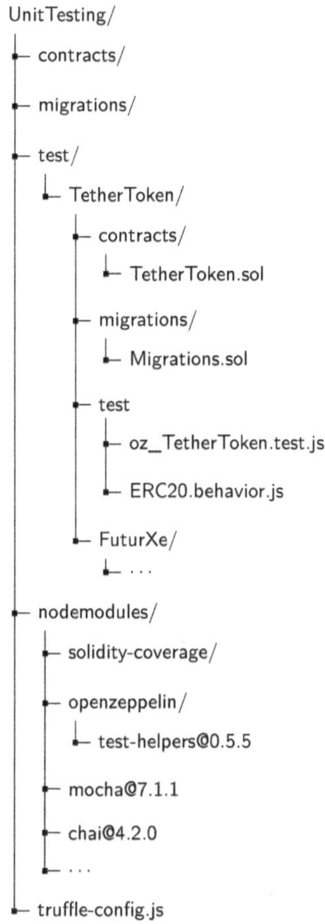

Figure 4.3: Truffle project environment

4.2.2 Test execution

Truffle allows to use JavaScript for testing, leveraging the Mocha testing framework to write more complex tests. One example is the contract() function (Listing 4.7, line 1). This function works exactly like Mocha's describe() function except it enables Truffle's clean-room features. Before each contract() function is run, the token contract is redeployed to the local blockchain so the tests within it run with a clean contract state. However, for each test file there is only one contract() function, meaning that testing isolation is accomplished by the setup method beforeEach(async function ()) alone (Listing 4.7, line 3). This method runs before each test block and deploys a new token contract instance to the blockchain, thus ensuring that upcoming tests will run under the same conditions. Additionally, contract() function also provides a list of

Listing 4.5: Truffle configuration file

```
module.exports = {
    networks: {},
    mocha: {},
    plugins: ["solidity-coverage"],

    // Configure your compilers
    compilers: {
        solc: {
            version: "0.4.17",    // Fetch exact version from solc-bin (
                default: truffle's version)
        }
    }
}
```

Listing 4.6: npm package file

```
{
  "name": "unittest",
  "version": "1.0.0",
  "description": "",
  "main": "index.js",
  "scripts": {
    "test": "echo \"Error: no test specified\" && exit 1"
  },
  "keywords": [],
  "author": "",
  "license": "ISC",
  "devDependencies": {
    "@openzeppelin/contracts": "^2.5.0",
    "@openzeppelin/test-helpers": "^0.5.5",
    "chai": "^4.2.0",
    "mocha": "^7.1.1",
    "solidity-coverage": "^0.7.5",
    "truffle": "^5.1.29"
  }
}
```

Ethereum accounts made available by **ganache-cli**.

Unit tests execute as follows using Truffle:

1. A set of external modules are imported.

2. The token's ABI is loaded from an artifact object provided by Truffle.

3. A **contract**() function is initialized providing access to the local ganache accounts and

defining the beforeEach(async function()) fixture function.

4. Before each test block is executed the beforeEach() takes place and deploys a fresh instance of the token contract to the blockchain

5. Each test block issues a set of transactions and assertions are conducted to attest if the contract's variables have the expected values.

6. Steps 3 to 5 are repeated for the remaining test blocks.

4.2.3 OpenZeppelin

A test in OpenZeppelin consists of two files, where the main test file (Listing 4.8) imports the second ERC20.behavior.js (line 9) and other third-party modules like Chai (line 2). Then, it loads the token contract object provided by Truffle (line 11) and defines the testing setup method (line 16), where the deployment of the testing token contract takes place and its variables are initialized. This will ensure that tests will execute under the same conditions every time. Finally, a set of unit tests (line 21) is called with contract related variables as arguments, which will be used to issue transactions.

The second file consists of the testing suite itself and this is where the actual tests are imported from. This file has a series of imported modules and functions that test particular behaviours of an ERC20 token like transfers or approvals. Those are the following.

- shouldBehaveLikeERC20

- shouldBehaveLikeERC20Transfer

- shouldBehaveLikeERC20Approve

For each one of the previous functions, a set of unit tests exist and can be identified by the function it (' testing a simple transfer ', async function ()). It is also worth mentioning that tests of no interest are commented. This is because Openzepelin often makes decisions on how tokens should behave with regards to particular scenarios. An example of this nature is the

Listing 4.7: Truffle test isolation

```
1  contract('Token', function ([_, initialHolder, recipient, anotherAccount]) {
2    const initialSupply = new BN('100');
3    beforeEach(async function () {
4      this.token = await Token.new(100,"TokenName","TKN",6,{'from':initialHolder
          });
5      this.token.initialSupply = initialSupply;
6    });
```

Listing 4.8: OpenZeppelin oz_Token.test.js test file

```javascript
1  const { BN, constants, expectEvent, expectRevert } = require('@openzeppelin/
       test-helpers');
2  const { expect } = require('chai');
3  const { ZERO_ADDRESS } = constants;
4
5  const {
6    shouldBehaveLikeERC20,
7    shouldBehaveLikeERC20Transfer,
8    shouldBehaveLikeERC20Approve,
9  } = require('./ERC20.behavior');
10
11 const Token = artifacts.require('Token');
12
13 contract('Token', function ([_, initialHolder, recipient, anotherAccount]) {
14   const initialSupply = new BN('100');
15
16   beforeEach(async function () {
17     this.token = await Token.new(100,"TokenName","TKN",8,{'from':initialHolder
           });
18     this.token.initialSupply = initialSupply;
19   });
20
21   shouldBehaveLikeERC20('ERC20', initialSupply, initialHolder, recipient,
         anotherAccount);
```

scenario where transactions with the zero address (0x0) as an argument should be reverted. One could argue that this is reasonable justified since zero address is often used and interpreted by the EVM as a deployment transaction (that is, to deploy a contract to the blockchain). However, ERC-20 does not make any remarks with regards to this scenario, and for that reason any test targeting this scenario is suppressed. Also, any test targeting the behavior of the mint() function will be left out of the testing suite as it is not considered to be in the scope of ERC-20 core.

4.2.4   Consensys

Similarly to OpenZeppelin, a test in Consensys consists of two files, where the main test file (Listing 4.9) imports the second file assertRevert.js (line 1). Then, the token contract object provided by Truffle is loaded (line 2) and a set of token related variables is defined (line 7,8,9 and 12). Next, the testing setup method is defined (line 15) and two instances of the same token contract are deployed to be used in upcoming tests (line 16 and 17).

Unlike Openzeppelin test file, Consensys defines its entire test suite within the main test file and uses assertRevert.js file as a module to handle event assertion. Once again, each test can be identified by the use of the function it('testing a simple transfer', async function () {...}).

Listing 4.9: Consensys **test_Token.js** test file

```javascript
const { assertRevert } = require('./assertRevert.js');
const INT = artifacts.require('TOKEN');

let HST;
let HST2;
const contract2test = TKN
const contract_name = 'TokenName';  ;
const contract_symbol = "TKN";
const contract_decimals = 6;
/*Try to test for overflows*/
const maxtokens = '
    115792089237316195423570985008687907853269984665640564039457584007913129639935
    ';
const initialBalance = 100;

contract(contract_name, (accounts) => {
  beforeEach(async () => {
    HST = await contract2test.new(100,"TokenName","TKN",6{"from": accounts
        [0]});
    HST2 = await contract2test.new(100,"TokenName","TKN",6{"from": accounts
        [0]});
  });
```

Chapter 5

Evaluation

This chapter provides an evaluation of the ERC-20 PBT framework using 10 contracts, comprising 8 real-world contracts plus the 2 reference implementations, Consensys and OpenZeppelin. We start by describing the evaluation setup in terms of contracts and test environment (Section 5.1). We then provide results concerning bug findings in these contracts (Section 5.2), a detailed performance analysis (Section 5.3), a comparison of our approach with the results of executing the Consensys and OpenZeppelin unit testing suites (Section 5.4), and, finally, bug findings for ERC-20 extensions regarding token minting, burning, and sale (Section 5.5).

5.1 Setup

5.1.1 Contracts tested

For our evaluation we considered the ERC-20 contract implementations listed in Table 5.1, all written in Solidity. The set includes:

- Four of the contracts listed in the top 10 ERC-20 implementations by the EtherScan site [22] as of September 2020: TetherToken, BNB, LinkToken (also known as ChainLink Token), and HBToken (also known as HuobiToken) – as shown in the table, these contracts have a high number of transactions recorded in the Ethereum blockchain;

- Four other real-word contracts that are referenced in an online collection of ERC-20 bug vulnerabilities from SecBit Labs [29]: BitAseanToken, FuturXe, INT, and SwiftCoin;

- The reference implementations of ERC-20: Consensys [7] and OpenZeppelin [40].

The source code of the contracts were collected from EtherScan for the real-world contracts, and the official GitHub repositories of Consensys and OpenZeppelin.

Table 5.1: Contracts used for evaluation.

	Contract	LOC	# Transactions
	BitAseanToken	153	≈ 2,000
	BNB	150	≈ 838,000
Real-world contracts	FutureXe	165	≈ 11,000
	HBToken	127	≈ 433,000
	INT (Internet Node Token)	184	≈ 126,000
	LinkToken	299	≈ 1,863,000
	SwiftCoin	175	≈ 78,000
	TetherToken	451	≈ 53,000,000
Reference implementations	Consensys	122	N/A
	OpenZeppelin	713	N/A

5.1.2 Environment

To derive the execution results, we made use of a dedicated Ubuntu 20.04 LTS machine hosted on Google Cloud Engine with 1 Intel Haswell CPU and 3.75 GB of RAM. Table 5.2 lists the software versions of the main common components (Linux kernel, Python, Node.js, and Ganache), as well as those specifically required by PBT (Hypothesis and Brownie) or the execution of the unit testing suites (Truffle along with a specific Solidity compiler; Brownie dynamically downloads and maintains the correct Solidity environment that is indicated in the source code of contracts).

Table 5.2: Software versions in the evaluation environment.

Software	Version
Linux kernel	5.4.0
Python	3.8.2
Node.js	10.19.0
Ganache	6.9.1
Hypothesis	5.23.7
Brownie	1.10.4
Truffle	5.1.29
Solidity compiler	0.4.17

5.2 ERC-20 bug findings

5.2.1 Methodology

We verified contracts using two different parameterisations for the pbt tool:

1. A generation of 100 examples while enabled verifications for event firing and function return values (pbt −n 100 −E −R ...);

2. And a generation of 1000 examples with disabled verifications for event firing and method return values (pbt −n 1000 ...).

The motivation for the two configurations is that bugs due to absent events or absent/invalid return values are easy to expose, typically with a simple function invocation. However, their reporting inhibits other assertions to be checked and bugs to be observed in the execution of a rule sequence.

The two configurations otherwise use the defaults for other pbt parameters meaning that shrinking is disabled (executions that make use of shrinking in are discussed in Section 5.3), the seed for test case generation is 0, and the stateful step count is 10.

5.2.2   Results

An overall summary of the bugs found is given in Table 5.3, and the cause of each bug is summarized per contract in Table 5.4, and the details of each bug can be found in Appendix B.

Table 5.3: Summary of results per bug category.

Category	# Bugs	# Contracts	
Absent event	4	4	BitAseanToken, BNB, INT, SwiftCoin
Absent return value	7	5	BitAseanToken, BNB, INT, SwiftCoin, TetherToken
Absent revert	8	3	FuturXe, HBToken, TetherToken
Invalid operation allowed	1	1	FuturXe
Operation not allowed	8	4	BNB, INT, FuturXe, TetherToken
Total	28	8	All except Consensys and OpenZeppelin

The bugs are grouped in the following categories:

- **Absent return value**: valid operations are executed, but the function at stake has a void return instead of returning a boolean true value;

- **Absent event**: valid operations are executed, but the function at stake fails to emit an expected event;

- **Absent revert**: an invalid operation does not proceed, but the function at stake fails to revert the transaction, instead using other means (false return values or assertion errors) to signal the invalid operation;

- **Operation not allowed**: a valid operation is not allowed to proceed, normally due to a bug in a pre-condition check and resulting in a revert;

- **Invalid operation allowed**: an invalid operation is allowed to proceed and potentially compromises the contract's state rather than reverting;

Table 5.4: ERC-20 bugs found.

Contract	Function	Bug type	Cause
BitAseanToken	approve	Absent event	Approval not fired for a valid approval.
	transfer	Absent return value	Fails to return a value for a valid transfer.
BNB	approve	Absent event	Approval not fired for a valid approval.
	approve	Operation not allowed	Reverts for an approval of 0 tokens.
	transfer	Absent return value	Fails to return a value for a valid transfer.
	transfer	Operation not allowed	Reverts for a transfer of 0 tokens.
	transferFrom	Operation not allowed	Reverts for a transfer of 0 tokens.
FuturXe	transfer	Absent revert	Returns false instead of reverting.
	transferFrom	Operation not allowed	Does not allow valid transfer.
	transferFrom	Invalid operation allowed	Allows invalid transfer and corrupts account balances and/or allowances.
	transferFrom	Absent revert	Returns false instead of reverting.
HBToken	transfer	Absent revert	Returns false instead of reverting.
	transferFrom	Absent revert	Returns false instead of reverting.
INT	approve	Absent event	Approval not fired for a valid approval.
	transfer	Absent return value	Fails to return a value for a valid transfer.
	transfer	Operation not allowed	Reverts when balance(owner) = amount.
	transferFrom	Operation not allowed	Reverts when balance(owner) = amount.
	transferFrom	Operation not allowed	Reverts when allowance(owner,msg.sender) = amount.
LinkToken	transfer	Absent revert	Assertion error instead of revert.
	transferFrom	Absent revert	Assertion error instead of revert.
SwiftCoin	approve	Absent event	Approval not fired for a valid approval.
	transfer	Absent return value	Fails to return a value for a valid transfer.
TetherToken	approve	Absent return value	Fails to return a value for a valid approval.
	approve	Operation not allowed	Allowance needs to be reset to 0 before new approval.
	transfer	Absent return value	Fails to return a value for a valid transfer.
	transfer	Absent revert	Assertion error instead of revert.
	transferFrom	Absent return value	Fails to return a value for a valid transfer.
	transferFrom	Absent revert	Assertion error instead of revert.

The last two categories can be considered the most serious, given that the contract's state in terms of account balances or allowances will differ from expected after a function invocation, either by having an faulty effect (when invalid operations are allowed to change state) or none (when valid operations are not allowed). In comparison, for the first three categories (absence of return values, events, or transaction reverts), the external interface with the contract is compromised but not the internal contract state.

5.3 Performance analysis

5.3.1 Methodology

The results in the previous section related to the generation of 1000 test examples, without shrinking enabled, and no details were given for the execution time and code coverage for the contracts. For a detailed performance analysis we additionally measured per contract:

- the use of 10, 25, and 100 examples and the corresponding measures for bug count, code coverage, and execution time;

- and the maximum length of falsifying examples for contracts that had bugs reported with and without shrinking, and the execution time overhead of shrinking.

The execution times we report are for coverage disabled, as coverage monitoring represents a significant overhead for execution (roughly up to 7 times slower). For coverage, we executed pbt −C ... to enable the calculation of Ethererum bytecode coverage by brownie. The measures are also given for executions with the pbt options for verification of return values and events disabled, hence the reported bug counts do not include these types of bugs.

5.3.2 Results

The results for bugs found, coverage, and execution time, are provided in aggregated form in Table 5.5 and per contract in Table 5.6. In the aggregated results, we may observe as expected that bug counts, coverage and execution times tend to grow with the number of examples.

Table 5.5: Performance analysis – aggregated results for bugs, code coverage and execution time.

# Examples	# Bugs	Coverage (%)	Time (s)
10	13	50	12 (1.20)
25	14	52	25 (1.00)
100	16	54	91 (0.91)
1000	17	54	1000 (1.00)

In more detail:

- The bug count grows from 13 to 17 as the number of examples grow, and in particular from 16 to 17 when the number of examples grows from 100 to 1000. The extra bug in this case is a 0-token transfer disallowed in transferFrom. For other seed values, we observed that this bug is already exposed for 100 examples.

- The coverage is highest in all cases for the Consensys and HBToken contracts, as they implement only the base ERC-20 functionality, while all others contracts have extra

Table 5.6: Performance analysis – bugs, code coverage and execution time per contract.

Contract	# Examples	# Bugs	Coverage (%)	Time (s)
BitAseanToken	10	0	51	9 (0.90)
	25	0	51	23 (0.92)
	100	0	56	99 (0.99)
	1000	0	56	1028 (1.03)
BNB	10	2	41	13 (1.30)
	25	2	43	24 (0.96)
	100	2	43	78 (0.78)
	1000	3	46	963 (0.96)
Consensys	10	0	88	9 (0.90)
	25	0	88	23 (0.92)
	100	0	94	81 (0.81)
	1000	0	94	1023 (1.02)
FuturXe	10	3	48	12 (1.20)
	25	4	63	22 (0.88)
	100	4	63	78 (0.78)
	1000	4	63	858 (0.86)
HBToken	10	2	88	13 (1.30)
	25	2	88	29 (1.16)
	100	2	88	98 (0.98)
	1000	2	88	1007 (1.01)
INT	10	2	30	14 (1.40)
	25	2	35	21 (0.84)
	100	3	36	77 (0.77)
	1000	3	36	927 (0.93)
LinkToken	10	2	30	13 (1.30)
	25	2	30	28 (1.12)
	100	2	30	102 (1.02)
	1000	2	30	1074 (1.07)
OpenZeppelin	10	0	54	9 (0.90)
	25	0	54	24 (0.96)
	100	0	54	106 (1.06)
	1000	0	54	1015 (1.02)
SwiftCoin	10	0	36	9 (0.90)
	25	0	36	23 (0.92)
	100	0	40	101 (1.01)
	1000	0	40	1068 (1.07)
TetherToken	10	2	35	14 (1.40)
	25	2	35	28 (1.12)
	100	3	35	88 (0.88)
	1000	3	35	1036 (1.04)

functionality (for token minting, burning, or sale, among others). From 100 to 1000 examples, the coverage is the same for all contracts except in the case of BNB where coverage grows from 43 % to 46%, in direct relation to the extra bug detected in BNB discussed above.

- The execution time tends to scales linearly with the number of examples. The execution time per example, shown in parenthesis in the tables, tends to converge to approximately 1 second.

The results for shrinking are provided in aggregated form in Table 5.7 and per contract in Table 5.8. In the tables we list the maximum number of rules used by a falsifying example with and without shrinking, the execution time when shrinking is enabled, and the overhead of execution time due to shrinking compared to executions where shrinking is disabled.

Table 5.7: Performance analysis – aggregated results for shrinking.

| | Max. rules/bug | | | |
# Examples	No sh.	With sh.	Avg. Time (s)	Avg. Overhead (%)
10	5	1	34	183
25	1	1	45	80
100	2	2	114	25
1000	2	2	990	-1

Listing 5.1: Falsifying example – INT bug exposed with 5 rules.

```
state = BrownieStateMachine ()
state.rule_transferFrom (st_amount=14975637542055411663, ...)
state.rule_transferFrom (st_amount=12238, ...)
state.rule_transferFrom (st_amount=712, ...)
state.rule_transferFrom (st_amount=40955, ...)
state.rule_approveAndTransferAll (st_owner=<Account '0
    x0063046686E46Dc6F15918b61AE2B121458534a5'>, st_receiver=<Account '0
    x0063046686E46Dc6F15918b61AE2B121458534a5'>, st_spender=<Account '0
    x0063046686E46Dc6F15918b61AE2B121458534a5'>)
state.teardown ()
...
  File "/home/edrdo/brownie_test/erc20_pbt.py", line 76, in rule_transferFrom
    tx = self.contract.transferFrom (
brownie.exceptions.VirtualMachineError: revert
...-
  File "contracts/INT.sol", line 67, in token.transferFrom:
...
        require (_value < allowance [_from][msg.sender]);
```

The main observations are as follows:

- Shrinking is effective in reducing the maximum number of rules used in a falsifying example

Table 5.8: Performance analysis – shrinking results per contract.

| Contract | # Examples | Max. rules/bug | | Time (s) | Overhead (%) |
		No sh.	With sh.		
BNB	10	2	1	23	77
	25	1	1	31	29
	100	1	1	85	9
	1000	1	1	968	1
FuturXe	10	3	1	38	217
	25	1	1	55	150
	100	1	1	125	60
	1000	1	1	859	0
HBToken	10	1	1	32	146
	25	1	1	42	45
	100	1	1	115	17
	1000	1	1	1001	-1
INT	10	5	1	32	129
	25	1	1	26	24
	100	1	1	79	3
	1000	1	1	887	-4
LinkToken	10	2	1	39	200
	25	1	1	56	100
	100	1	1	117	15
	1000	1	1	1104	3
TetherToken	10	2	1	41	193
	25	1	1	62	121
	100	2	2	163	85
	1000	2	2	1123	8

when the number of examples is only 10. For 25, 100, and 1000 examples there is no difference in such a metric. The maximum number of rules of 5 is observed for 10 examples and an INT bug as shown in Listing 5.1. The corresponding falsifying example when shrinking is enabled, shown in Listing 5.2, uses just one rule even if it is the rule_approveAndTransferAll composed rule that invokes rule_approve and rule_transferFrom in sequence. Along with rule_approveAndTransferAll, this rule exercises boundary conditions explicitly.

- All bugs found can be reproduced with just 1 rule except for a bug in TetherToken that uses 2 rules, found when the number of examples is set to 100 and 1000. This contract requires an allowance to be reset to 0 using approve before new a new value is set using the same function. The falsifying examples without shrinking and with shrinking are shown respectively in Listing 5.3 and Listing 5.4. Even though rule_approve is used twice in both cases, shrinking finds an input value of 1 rather than 1024, leading to a more understandable falsifying example,. In many cases, shrinking finds more adequate input values for rule invocation.

Listing 5.2: Falsifying example – INT bug exposed with 1 rule after shrinking.

```
Falsifying example:
state = BrownieStateMachine()
state.rule_approveAndTransferAll(st_owner=<Account '0
    x66aB6D9362d4F35596279692F0251Db635165871'>, st_receiver=<Account '0
    x66aB6D9362d4F35596279692F0251Db635165871'>, st_spender=<Account '0
    x66aB6D9362d4F35596279692F0251Db635165871'>)
state.teardown()
...
  File "/home/edrdo/brownie_test/erc20_pbt.py", line 76, in rule_transferFrom
    tx = self.contract.transferFrom(
brownie.exceptions.VirtualMachineError: revert
...
  File "contracts/INT.sol", line 67, in token.transferFrom:
...
      require (_value < allowance[_from][msg.sender]);
```

Listing 5.3: Falsifying example – Tether token bug exposed with 2 rules and shrinking disabled.

```
Falsifying example:
state = BrownieStateMachine()
state.rule_approve(st_amount=1024, st_owner=<Account '0
    x33A4622B82D4c04a53e170c638B944ce27cffce3'>, st_spender=<Account '0
    x33A4622B82D4c04a53e170c638B944ce27cffce3'>)
state.rule_approve(st_amount=1024, st_owner=<Account '0
    x33A4622B82D4c04a53e170c638B944ce27cffce3'>, st_spender=<Account '0
    x33A4622B82D4c04a53e170c638B944ce27cffce3'>)
state.teardown()
...
brownie.exceptions.VirtualMachineError: revert
  File "contracts/TetherToken.sol", line 205, in StandardToken.approve:
...
      require(!((_value != 0) && (allowed[msg.sender][_spender] != 0)));
```

- Shrinking happens only after raw example generation. As more examples are generated, Hypothesis internally replaces the falsifying example for an assertion error if it finds to be shorter in the number of invoked rules. Thus, if more examples are generated, the subsequent shrinking effort tends to be lower. The results show that the execution time overhead of using shrinking progressively decreases as the number of generated examples grow. For 1000 examples we found little or no overhead due to shrinking.

Listing 5.4: Falsifying example – Tether token bug exposed with 2 rules and shrinking enabled.

```
Falsifying example:
state = BrownieStateMachine()
state.rule_approve(st_amount=1, st_owner=<Account '0
    x33A4622B82D4c04a53e170c638B944ce27cffce3'>, st_spender=<Account '0
    x33A4622B82D4c04a53e170c638B944ce27cffce3'>)
state.rule_approve(st_amount=1, st_owner=<Account '0
    x33A4622B82D4c04a53e170c638B944ce27cffce3'>, st_spender=<Account '0
    x33A4622B82D4c04a53e170c638B944ce27cffce3'>)
state.teardown()
...
brownie.exceptions.VirtualMachineError: revert
  File "contracts/TetherToken.sol", line 205, in StandardToken.approve:
...
    require(!((_value != 0) && (allowed[msg.sender][_spender] != 0)));
```

5.4 Comparison to unit testing

5.4.1 Methodology

We compared the PBT approach with unit testing suites of the Consensys and OpenZeppelin reference implementations. We measured the efficiency of the unit testing suites and pbt in terms of bugs found, execution time, and coverage for Ethereum bytecode. The two unit testing suites (organised as described in Chapter 4) were executed for every contract, and the results were compared with the invocation of the pbt −n 1000 reported in earlier sections. As for pbt (in Section 5.3), the execution times we report are for coverage disabled in the unit testing suites. Coverage monitoring again represents a significant overhead for execution time (roughly 1.5 times slower).

We did not enable verification of return values or events through the −E and −R switches of pbt for a fairer comparison. The motivation is that the unit testing suites do not verify return values at all and have specific tests that check only for event emission. On the other hand, event and return value verification limits the ability of pbt exposing other bugs. Given that events are not checked by pbt with these settings, the bug counts we report for unit testing suites do not include absent event bugs.

5.4.2 Results

The results are provided in aggregate form in Table 5.9, with an entry for our approach, and two other entries for the Consensys and OpenZeppelin testing suites. In the table we list: the lines of code (LOC) in our StateMachine class, and the unit testing suites; the number of tests executed for each approach; the number of bugs found; the average coverage per contract; and the execution time per test. For the number of bugs column, we report within parenthesis the total number of bugs found per each test driver including absent approval events (4 bugs are detectable by PBT, Consensys and OpenZeppelin) and absent return values (7 detectable by PBT, none by Consensys and OpenZeppelin). In Table 5.10 we provide results in detailed form per contract, regarding the number of bugs found, code coverage, and execution time per test.

Table 5.9: Comparison to unit testing – aggregated results.

Tests	LOC	# Tests	# Total Bugs	Avg. Coverage (%)	Time per test (s)
PBT	173	1000	17 (28)	54	0.99
Consensys	225	17	14 (18)	57	0.75
OpenZeppelin	410	25	15 (19)	55	0.54

Table 5.10: Comparison to unit testing – results per contract.

B: bugs found   **C**: coverage (%)   **T**: execution time per test (s)

Contract	PBT			Consensys			OpenZeppelin		
	B	C	T	B	C	T	B	C	T
BitAseanToken	0	56	1.03	0	41	0.76	0	43	0.48
BNB	3	43	0.96	2	54	0.82	1	54	0.48
Consensys	0	94	1.02	0	100	0.70	0	100	0.48
FuturXe	4	63	0.86	4	65	0.70	4	67	0.48
HBToken	2	88	1.01	2	76	0.70	2	76	0.48
INT	3	36	0.93	2	36	0.76	3	22	0.52
LinkToken	2	30	1.01	2	51	0.76	2	51	0.72
OpenZeppelin	0	54	1.01	0	49	0.70	0	46	0.56
SwiftCoin	0	40	1.06	0	39	0.76	0	39	0.48
TetherToken	3	35	1.03	2	54	0.82	3	54	0.68

The results overall indicate that pbt was able to expose 17 bugs in total, 3 more bugs than the Consensys test suite and 2 more than the OpenZeppelin test suite, relative increases of 21 % and 13 % respectively. Per contract, we see that more bugs were exposed for BNB, INT and TetherToken.

In detail, the difference in bug counts can be explained as follows:

- For BNB, pbt finds 3 bugs, all related to the fact that an approval or transfer of 0 tokens reverts in approve, transfer, and transferFrom. The bug in transferFrom is not exposed by Consensys or OpenZeppelin, and the bug in approve is not exposed by OpenZeppelin.

- For INT, Consensys fails to detect the bug that transferFrom() reverts for a valid transfer when allowance(owner,msg.sender) = amount.

- For TetherToken, Consensys fails to detect that an allowance needs to be reset to 0 before a new approval.

Regarding coverage, even if the numbers are derived differently and correspond to different coverage metrics for pbt and the unit testing suites, the average coverage happens to be very similar for all three test drivers. Moreover, the numbers tend to be also very similar for most of the contracts. The coverage is only significantly lower for pbt in the case of LinkToken and TetherToken. Finally, regarding execution time, it is clear that pbt takes more time per test, roughly 1 second per test on average, 32 % more than the Consensys test suite, and 83 % more than the OpenZeppelin test suite.

5.5 ERC-20 extensions

We now provide results regarding ERC-20 extensions described earlier in Chapter 3, and bugs found using the corresponding extensions to the ERC-20 StateMachine: MintingStateMachine, BurningStateMachine, and BuySellStateMachine.

5.5.1 Token minting

In the set of contracts we considered for testing, token minting is supported by BitAseanToken, FuturXe, INT, OpenZeppelin, and SwiftCoin. Recall that a call to mintToken(addr,amount) generates new amount tokens and associates these tokens to the addr account, thus the contract's total supply of tokens and the balance of addr must be increased by amount. These updates need to be checked for overflow, i.e., the corresponding sums must not exceed $2^{256} - 1$, otherwise invalid values will result for the total supply of tokens and/or account balances. We found bugs due to overflow errors in all of the contracts with minting support, except OpenZeppelin.

In Listing 5.5 we illustrate the bug for BitAseanToken, and the other offending contracts (FuturXe, INT, SwiftCoin) have a similar implementation. As shown in lines 2 and 3, two unchecked sums are used to update the contract's state. The output for a corresponding falsifying example is shown in Listing 5.6, illustrating that a call to mintToken does not revert in case of an overflow and that the token's total supply value becomes invalid. In contrast, the bugs is avoided by the OpenZeppelin implementation shown in Listing 5.7. In this case, overflows are detected through calls to SafeMath.add() function that reverts the transaction in case an overflow is detected.

Listing 5.5: Code for mintToken() in BitAseanToken.

```
1  function mintToken(address target, uint256 mintedAmount) onlyOwner {
2      balanceOf[target] += mintedAmount;  // unchecked for overflow
3      totalSupply += mintedAmount;        // unchecked for overflow
4      Transfer(0, this, mintedAmount);
5      Transfer(this, target, mintedAmount);
6  }
```

Listing 5.6: Falsifying example – mintToken() overflow bug in BitAseanToken.

```
Falsifying example:
state = BrownieStateMachine()
state.rule_mint(st_amount
    =87489218939561473620073929897326909490865436869672301917939527265232132906863,
     st_receiver=<Account '0x0063046686E46Dc6F15918b61AE2B121458534a5'>)
state.rule_mint(st_amount
    =87489218939561473620073929897326909490865436869672301917939527265232132906863,
     st_receiver=<Account '0x0063046686E46Dc6F15918b61AE2B121458534a5'>)
state.teardown()
...
AssertionError: Transaction did not revert
...
During handling of the above exception, another exception occurred:
AssertionError: totalSupply() : expected value
    87489218939561473620073929897326909490865436869672301917939527265232132907863,
    actual value was
    59186348641806751816576874785965911128460889073704039796421470522551136174790
```

Listing 5.7: Code for mintToken() in OpenZeppelin.

```solidity
 1  function _mint(address account, uint256 amount) internal {
 2    require(account != address(0), "ERC20: mint to the zero address");
 3    _totalSupply = _totalSupply.add(amount);                    // checked for overflow
 4    _balances[account] = _balances[account].add(amount); // checked for overflow
 5    emit Transfer(address(0), account, amount);
 6  }
 7  function mintToken(address account, uint256 amount)
 8    public onlyMinter returns (bool) {
 9    _mint(account, amount);
10    return true;
11  }
```

5.5.2 Token burning

In the set of contracts we considered for testing, token burning is supported by BNB and
INT (INT is the only contract that supports both minting and burning). Recall that a call to
burn(amount) destroys amount tokens held by the caller (msg.sender), thus the contract's total
supply of tokens and the balance of the caller are decreased by amount. This would be prone
to an underflow if amount exceeds the caller's balance, but both contracts at stake revert the
transaction when balanceOf(msg.sender) < amount. Unlike in the case of token minting, no bugs
are observed in regard to token total supply or account balances. The necessary verification is
illustrated in line 2 of the code for burn in BNB, shown in Listing 5.8, In the same code we may

also observe in any case the convention of using SafeMath.sub performing the subtractions to check for underflows. In contrast, the subtractions are unchecked for INT, as shown in Listing 5.9, but underflows are prevented as well.

Listing 5.8: Code for burn() in BNB.

```
1 function burn(uint256 _value) returns (bool success) {
2   if (balanceOf[msg.sender] < _value) throw;
3   if (_value <= 0) throw;
4   balanceOf[msg.sender] = SafeMath.safeSub(balanceOf[msg.sender], _value);
5   totalSupply = SafeMath.safeSub(totalSupply, _value);
6   Burn(msg.sender, _value);
7   return true;
8 }
```

Listing 5.9: Code for burn() in INT.

```
1 function burn(uint256 _value) returns (bool success) {
2   require (balanceOf[msg.sender] > _value);
3   balanceOf[msg.sender] -= _value;
4   totalSupply -= _value;
5   Burn(msg.sender, _value);
6   return true;
7 }
```

The modelling assumptions in BurningStateMachine allow the case of burning 0 tokens (like a void 0-token transfer is allowed in ERC-20). The BNB token does not allow it, similarly to bugs we found in transfer and transferFrom for the same contract, due the explicit check in line 3 of Listing 5.8, as illustrated by the falsifying example in Listing 5.10. The INT contract also has a bug: in line 2 of Listing 5.9, require (balanceOf[msg.sender] > _value) does not allow for the sender to burn all of its tokens, reverting when balanceOf[msg.sender] == _value, as illustrated by the falsifying example in Listing 5.11.

5.5.3   Token sale

In the set of contracts we considered for testing, token sale is supported by the INT and SwiftToken. In these contracts, buy and sell allow the sender to buy or sell tokens in exchange of ether, and setPrices is used to set the buy and sell prices of the token in ether.

In both contracts, setPrices allows both buy and sell prices to be to 0, and these prices are in fact initially set to 0 by default. To allow for easier reproduction of bugs BurningStateMachine sets both prices to 1 during setup, but rule_setPrices may set back the prices to 0. Buying tokens should not be allowed for a buy price of 0 but a buy call does not revert in this case,

Listing 5.10: Falsifying example for burn() bug in BNB.

```
Falsifying example:
state = BrownieStateMachine()
state.rule_burn_all(st_sender=<Account '0
    x33A4622B82D4c04a53e170c638B944ce27cffce3'>)
state.teardown()
...
brownie.exceptions.VirtualMachineError: revert
Trace step -1, program counter 1598:
  File "contracts/BNB.sol", line 116, in BNB.burn:
...

      if (_value <= 0) throw;
```

Listing 5.11: Falsifying example for burn() bug in INT.

```
state = BrownieStateMachine()
state.rule_burn_all(st_sender=<Account '0
    x66aB6D9362d4F35596279692F0251Db635165871'>)
state.teardown()
...
brownie.exceptions.VirtualMachineError: revert
...
  File "contracts/INT.sol", line 98, in token.burn:
  ...

        require (balanceOf[msg.sender] > _value);
```

and instead a division-by-zero occurs in both contracts. The issue is illustrated for SwiftCoin in terms of the code in Listing 5.12, and the falsifying example in Listing 5.13.

Another bugs arises but only for INT and the sell () operation. Similarly to transfer and transferFrom in the same contract, sell (amount) reverts when balanceOf(msg.sender) == amount, as illustrated by the falsifying example in Listing 5.14.

Listing 5.12: Code for buy() in SwiftCoin.

```
1 function buy() payable {
2     uint amount = msg.value / buyPrice; // Possible division by zero
3     if (balanceOf[this] < amount) throw;
4     balanceOf[msg.sender] += amount;
5     balanceOf[this] -= amount;
6     Transfer(this, msg.sender, amount);
7 }
```

Listing 5.13: Falsifying example for divison-by-zero bug in SwiftCoin.

```
Falsifying example:
state = BrownieStateMachine()
state.rule_setPrices(st_amount=0)
state.rule_buy(st_amount=0, st_sender=<Account '0
   x66aB6D9362d4F35596279692F0251Db635165871'>)
state.teardown()
...
brownie.exceptions.VirtualMachineError: invalid opcode: Division by zero
  File "contracts/SwiftCoin.sol", line 158, in SwftCoin.buy:
   ...
       uint amount = msg.value / buyPrice;
```

Listing 5.14: Falsifying example for total token sale in INT.

```
Falsifying example:
state = BrownieStateMachine()
state.rule_sellAll(st_sender=<Account '0
   x66aB6D9362d4F35596279692F0251Db635165871'>)
state.teardown()
...
brownie.exceptions.VirtualMachineError: revert
Trace step -1, program counter 2896:
  File "contracts/INT.sol", line 135, in INTToken._transfer:
...
       require (balanceOf[_from] > _value);
```

Chapter 6

Conclusions

6.1 Summary

In this book, we presented a property-based testing framework for ERC-20 contracts. We developed an ERC-20 model and some common extensions to it as rule-based state machines on top of the Brownie framework. We conducted an evaluation of this approach over 10 ERC-20 contracts written in the Solidity language, including 8 real-world contracts and the 2 reference implementations of ERC-20, with the following highlights:

- Bugs were found for all contracts except for the reference implementations. Since we considered some of the most widely used ERC-20 tokens like TetherToken, LinkToken, and BNB (the current top 3 tokens listed by Etherscan in terms of market capitalization [22]), this provides a strong suggestion that ERC-20 contracts commonly exhibit bugs and deviations to the standard.

- The PBT approach was able to expose more bugs than the unit testing suites from Consensys and OpenZeppelin, demonstrating the potential of PBT over unit testing.

- Additionally, we reported other kind of bugs such as divisions by zero and overflows for other types of functionality in contracts (token minting, burning and sale).

Given these findings, we generally conclude that PBT is a sound approach to expose bugs in ERC-20 contracts, and potentially other types of Ethereum smart contracts.

6.2 Future work

Future work concerns deeper analysis of experiments, tests and new adaptions to the presented testing framework. Hence, this work can take an interesting number of directions. We propose the following:

- A wider universe of ERC-20 contracts may be considered for evaluation. Etherscan reports 299,431 ERC-20 tokens in the Ethereum network! In this work, we only addressed 8 real-world contracts.

- The PBT approach may be used to address other token standards related to ERC-20. ERC-777 [12] is a direct extension of ERC-20, defining new ways of interacting with tokens. ERC-721 [17] defines a standard for non-fungible tokens, making it possible to represent any arbitrary data or asset as a unique token, drastically increasing the scope of what can be represented as tokens in the Ethereum blockchain. Lastly, ERC-1155 [42] defines standard interface for contracts that manage multiple token types, allowing a single contract to manage a combination of fungible or non-fungible tokens. This standard also features batch transfers, where multiple tokens can be sent in a single transaction, thus offering significant savings on gas costs.

- Rule-based state machines may be defined to test common ERC-20 token extensions, such as crowd-sale [9], pausing [19] and migration [18] to name a few.

- Property-based testing has an interesting variation called targeted property-based testing [30] that is supported by Hypothesis. In a nutshell, instead of being completely random, this approach uses a search-based component to guide the input generation towards values that have a higher probability of falsifying a property. This can prove to be a valid approach for testing smart contracts also.

Appendix A

Source code

This appendix lists the main source code of the PBT framework for ERC-20 contracts. We list the test logic in Python (Listing A.1) and the code for the pbt shell script (Listing A.2) The source code can also be found at GitHub [46].

Listing A.1: Test logic (erc20_pbt.py)

```python
import os
import brownie
from brownie.test import strategy
from brownie.exceptions import VirtualMachineError


class StateMachine:
    st_amount = strategy("uint256")
    st_owner = strategy("address")
    st_spender = strategy("address")
    st_sender = strategy("address")
    st_receiver = strategy("address")

    def __init__(self, accounts, contract, totalSupply, DEBUG=None):
        self.accounts = accounts
        self.contract = contract
        self.totalSupply = totalSupply
        self.DEBUG = DEBUG != None or os.getenv("PBT_DEBUG", "no") == "yes"
        self.VERIFY_EVENTS = os.getenv("PBT_VERIFY_EVENTS") == "yes"
        self.VERIFY_RETURN_VALUES = (
            os.getenv("PBT_VERIFY_RETURN_VALUES") == "yes"
        )

    def setup(self):
        if self.DEBUG:
            print("setup()")
        self.allowances = dict()
        self.balances = {i: 0 for i in self.accounts}
        self.balances[self.accounts[0]] = self.totalSupply
        self.value_failure = False

    def teardown(self):
        if self.DEBUG:
            print("teardown()")
        if not self.value_failure:
            self.verifyTotalSupply()
            self.verifyAllBalances()
            self.verifyAllAllowances()

    def rule_transfer(self, st_sender, st_receiver, st_amount):
        if self.DEBUG:
            print(
                "transfer({}, {}, {})".format(st_sender, st_receiver, st_amount)
            )
        if st_amount <= self.balances[st_sender]:
```

```python
46              with normal():
47                  tx = self.contract.transfer(
48                      st_receiver, st_amount, {"from": st_sender}
49                  )
50                  self.verifyTransfer(st_sender, st_receiver, st_amount)
51                  self.verifyEvent(
52                      tx,
53                      "Transfer",
54                      {"from": st_sender, "to": st_receiver, "value": st_amount},
55                  )
56                  self.verifyReturnValue(tx, True)
57          else:
58              with brownie.reverts():
59                  self.contract.transfer(
60                      st_receiver, st_amount, {"from": st_sender}
61                  )
62
63      def rule_transferFrom(self, st_spender, st_owner, st_receiver, st_amount):
64          if self.DEBUG:
65              print(
66                  "transferFrom({}, {}, {}, [from: {}])".format(
67                      st_owner, st_receiver, st_amount, st_spender
68                  )
69              )
70          if st_amount == 0 or (
71              (st_owner, st_spender) in self.allowances.keys()
72              and self.balances[st_owner] >= st_amount
73              and self.allowances[(st_owner, st_spender)] >= st_amount
74          ):
75              with normal():
76                  tx = self.contract.transferFrom(
77                      st_owner, st_receiver, st_amount, {"from": st_spender}
78                  )
79                  self.verifyTransfer(st_owner, st_receiver, st_amount)
80                  if st_amount != 0:
81                      self.verifyAllowance(st_owner, st_spender, -st_amount)
82                  self.verifyEvent(
83                      tx,
84                      "Transfer",
85                      {"from": st_owner, "to": st_receiver, "value": st_amount},
86                  )
87                  self.verifyReturnValue(tx, True)
88          else:
89              with brownie.reverts():
90                  self.contract.transferFrom(
91                      st_owner, st_receiver, st_amount, {"from": st_spender}
92                  )
93
94      def rule_approve(self, st_owner, st_spender, st_amount):
95          if self.DEBUG:
96              print("approve({}, {}, {})".format(st_owner, st_spender, st_amount))
97          with (normal()):
98              tx = self.contract.approve(
99                  st_spender, st_amount, {"from": st_owner}
100             )
101             self.verifyAllowance(st_owner, st_spender, st_amount)
102             self.verifyEvent(
103                 tx,
104                 "Approval",
105                 {"owner": st_owner, "spender": st_spender, "value": st_amount},
106             )
107             self.verifyReturnValue(tx, True)
108
109     def rule_transferAll(self, st_sender, st_receiver):
110         self.rule_transfer(st_sender, st_receiver, self.balances[st_sender])
111
112     def rule_approveAndTransferAll(self, st_owner, st_spender, st_receiver):
113         amount = self.balances[st_owner]
114         self.rule_approve(st_owner, st_spender, amount)
115         self.rule_transferFrom(st_spender, st_owner, st_receiver, amount)
116
117     def verifyTotalSupply(self):
118         self.verifyValue(
119             "totalSupply()", self.totalSupply, self.contract.totalSupply()
120         )
121
122     def verifyAllBalances(self):
123         for account in self.balances:
124             self.verifyBalance(account)
125
```

```python
126    def verifyAllAllowances(self):
127        for (owner, spender) in self.allowances:
128            self.verifyAllowance(owner, spender)
129
130    def verifyBalance(self, addr):
131        self.verifyValue(
132            "balanceOf({})".format(addr),
133            self.balances[addr],
134            self.contract.balanceOf(addr),
135        )
136
137    def verifyTransfer(self, src, dst, amount):
138        self.balances[src] -= amount
139        self.balances[dst] += amount
140        self.verifyBalance(src)
141        self.verifyBalance(dst)
142
143    def verifyAllowance(self, owner, spender, delta=None):
144        if delta != None:
145            if delta >= 0:
146                self.allowances[(owner, spender)] = delta
147            elif delta < 0:
148                self.allowances[(owner, spender)] += delta
149        self.verifyValue(
150            "allowance({},{})".format(owner, spender),
151            self.allowances[(owner, spender)],
152            self.contract.allowance(owner, spender),
153        )
154
155    def verifyReturnValue(self, tx, expected):
156        if self.VERIFY_RETURN_VALUES:
157            self.verifyValue("return value", expected, tx.return_value)
158
159    def verifyValue(self, msg, expected, actual):
160        if expected != actual:
161            self.value_failure = True
162            raise AssertionError(
163                "{} : expected value {}, actual value was {}".format(
164                    msg, expected, actual
165                )
166            )
167
168    def verifyEvent(self, tx, eventName, data):
169        if self.VERIFY_EVENTS:
170            if not eventName in tx.events:
171                raise AssertionError(
172                    "{}: event was not fired".format(eventName)
173                )
174            ev = tx.events[eventName]
175            for k in data:
176                if not k in ev:
177                    raise AssertionError(
178                        "{}.{}: absent event data".format(eventName, k)
179                    )
180                self.verifyValue("{}.{}".format(eventName, k), data[k], ev[k])
181
182
183 class MintingStateMachine(StateMachine):
184    def __init__(self, accounts, contract, totalSupply, DEBUG=None):
185        StateMachine.__init__(self, accounts, contract, totalSupply, DEBUG)
186
187    def rule_mint(self, st_receiver, st_amount):
188        if self.DEBUG:
189            print("mint({}, {})".format(st_receiver, st_amount))
190        if st_amount + self.totalSupply <= 2 ** 256 - 1:
191            with normal():
192                self.contract.mintToken(
193                    st_receiver, st_amount, {"from": self.accounts[0]}
194                )
195                self.totalSupply += st_amount
196                self.balances[st_receiver] += st_amount
197                self.verifyBalance(st_receiver)
198                self.verifyTotalSupply()
199        else:
200            with (brownie.reverts()):
201                self.contract.mintToken(
202                    st_receiver, st_amount, {"from": self.accounts[0]}
203                )
204
205
```

```python
class BurningStateMachine(StateMachine):
    def __init__(self, accounts, contract, totalSupply, DEBUG=None):
        StateMachine.__init__(self, accounts, contract, totalSupply, DEBUG)

    def rule_burn(self, st_sender, st_amount):
        if self.DEBUG:
            print("burn({}, {})".format(st_sender, st_amount))
        if st_amount >= 0 and self.balances[st_sender] >= st_amount:
            with normal():
                tx = self.contract.burn(st_amount, {"from": st_sender})
                self.totalSupply -= st_amount
                self.balances[st_sender] -= st_amount
                self.verifyBalance(st_sender)
                self.verifyTotalSupply()
                self.verifyEvent(
                    tx, "Burn", {"from": st_sender, "value": st_amount}
                )
        else:
            with (brownie.reverts()):
                self.contract.burn(st_amount, {"from": st_sender})

    def rule_burn_all(self, st_sender):
        self.rule_burn(st_sender, self.balances[st_sender])


class BuySellStateMachine(StateMachine):
    INITIAL_BUY_PRICE = 1
    INITIAL_SELL_PRICE = 1

    def __init__(self, accounts, contract, totalSupply, DEBUG=None):
        StateMachine.__init__(self, accounts, contract, totalSupply, DEBUG)

    def setup(self):
        # Base state machine setup
        StateMachine.setup(self)

        # Set initial prices
        self.buyPrice = self.INITIAL_BUY_PRICE
        self.sellPrice = self.INITIAL_SELL_PRICE

        # Set sell and buy price
        self.contract.setPrices(
            self.sellPrice, self.buyPrice, {"from": self.accounts[0]}
        )

        # Set up model for ether balance
        self.ethBalances = {i: i.balance() for i in self.accounts}
        self.ethBalances[self.contract] = self.contract.balance()

        # Model contract balance as well
        self.balances[self.contract] = 0

    def teardown(self):
        StateMachine.teardown(self)
        for x in self.ethBalances:
            self.verifyEthBalance(x)

    def rule_setPrices(self, st_amount):
        if self.DEBUG:
            print("setPrices({}, {})".format(st_amount, st_amount))
        self.buyPrice = st_amount
        self.sellPrice = st_amount
        self.contract.setPrices(
            self.sellPrice, self.buyPrice, {"from": self.accounts[0]}
        )
        self.verifyValue("buyPrice", self.buyPrice, self.contract.buyPrice())
        self.verifyValue("sellPrice", self.sellPrice, self.contract.sellPrice())

    def rule_sell(self, st_sender, st_amount):
        if self.DEBUG:
            print("sell({}, {})".format(st_sender, st_amount))
        ether = st_amount * self.sellPrice
        if (
            self.balances[st_sender] >= st_amount
            and self.ethBalances[self.contract] >= ether
        ):
            with normal():
                tx = self.contract.sell(st_amount, {"from": st_sender})
                self.verifySale(st_sender, self.contract, st_amount, ether, tx)
        else:
```

```python
286                with (brownie.reverts()):
287                    self.contract.sell(st_amount, {"from": st_sender})
288
289        def rule_buy(self, st_sender, st_amount):
290            if self.DEBUG:
291                print("buy({}, {})".format(st_sender, st_amount))
292            if (
293                self.buyPrice > 0
294                and self.ethBalances[st_sender] >= st_amount
295                and self.balances[self.contract] >= st_amount // self.buyPrice
296            ):
297                with normal():
298                    tx = self.contract.buy({"from": st_sender, "value": st_amount})
299                    self.verifySale(
300                        self.contract,
301                        st_sender,
302                        st_amount // self.buyPrice,
303                        st_amount,
304                        tx,
305                    )
306            elif self.ethBalances[st_sender] >= st_amount:
307                with (brownie.reverts()):
308                    self.contract.buy({"from": st_sender, "value": st_amount})
309
310        def rule_sellAll(self, st_sender):
311            self.rule_sell(st_sender, self.balances[st_sender])
312
313        def verifyEthBalance(self, addr):
314            self.verifyValue(
315                "ethBalance({})".format(addr),
316                self.ethBalances[addr],
317                addr.balance(),
318            )
319
320        def verifySale(self, a, b, tokens, ether, tx):
321            self.balances[a] -= tokens
322            self.balances[b] += tokens
323            self.ethBalances[a] += ether
324            self.ethBalances[b] -= ether
325            self.verifyBalance(a)
326            self.verifyBalance(b)
327            self.verifyEthBalance(a)
328            self.verifyEthBalance(b)
329            self.verifyEvent(tx, "Transfer", {"from": a, "to": b, "value": tokens})
330
331
332    def patch_hypothesis_for_seed_handling(seed):
333        import hypothesis
334
335        h_run_state_machine = hypothesis.stateful.run_state_machine_as_test
336
337        def run_state_machine(state_machine_factory, settings=None):
338            state_machine_factory._hypothesis_internal_use_seed = seed
339            h_run_state_machine(state_machine_factory, settings)
340
341        hypothesis.stateful.run_state_machine_as_test = run_state_machine
342
343
344    def patch_brownie_for_assertion_detection():
345        from brownie.test.managers.runner import RevertContextManager
346        from brownie.exceptions import VirtualMachineError
347
348        f = RevertContextManager.__exit__
349
350        def alt_exit(self, exc_type, exc_value, traceback):
351            if exc_type is VirtualMachineError:
352                exc_value.__traceback__.tb_next = None
353                if exc_value.revert_type != "revert":
354                    return False
355            return f(self, exc_type, exc_value, traceback)
356
357        RevertContextManager.__exit__ = alt_exit
358
359
360    def register_hypothesis_profiles():
361        import hypothesis
362        from hypothesis import settings, Verbosity, Phase
363
364        stateful_step_count = int(os.getenv("PBT_STATEFUL_STEP_COUNT", 10))
365        max_examples = int(os.getenv("PBT_MAX_EXAMPLES", 100))
```

```python
        derandomize = True
        seed = int(os.getenv("PBT_SEED", 0))

        if seed != 0:
            patch_hypothesis_for_seed_handling(seed)
            derandomize = False

        patch_brownie_for_assertion_detection()

        settings.register_profile(
            "generate",
            stateful_step_count=stateful_step_count,
            max_examples=max_examples,
            phases=[Phase.generate],
            report_multiple_bugs=True,
            derandomize=derandomize,
            print_blob=True,
        )

        settings.register_profile(
            "shrinking",
            stateful_step_count=stateful_step_count,
            max_examples=max_examples,
            phases=[Phase.generate, Phase.shrink],
            report_multiple_bugs=True,
            derandomize=derandomize,
            print_blob=True,
        )


class NoRevertContextManager:
    def __init__(self):
        pass

    def __enter__(self):
        pass

    def __exit__(self, exc_type, exc_value, traceback):
        if exc_type is None:
            return True
        import traceback

        if exc_type is VirtualMachineError:
            exc_value.__traceback__.tb_next = None
        elif exc_type is AssertionError:
            exc_value.__traceback__.tb_next = None
        return False


def normal():
    return NoRevertContextManager()
```

Listing A.2: PBT execution script (pbt)

```bash
1  #! /bin/bash
2
3  export PYTHONPATH=$(dirname $0)
4
5  usage() {
6    echo "Usage:"
7    echo " $(basename $0) [options] test1 ... testn"
8    echo "Options:"
9    echo " -c <arg> : set stateful step count"
10   echo " -n <arg> : set maximum examples"
11   echo " -s <arg> : set seed for tests"
12   echo " -C       : measure coverage"
13   echo " -D       : enable debug output"
14   echo " -E       : enable verification of events"
15   echo " -R       : enable verification of return values"
16   echo " -S       : enable shrinking"
17  }
18
19  PBT_DEBUG=no
20  PBT_SEED=0
21  PBT_MAX_EXAMPLES=100
22  PBT_STATEFUL_STEP_COUNT=10
23  PBT_PROFILE=generate
24  PBT_VERIFY_EVENTS=no
25  PBT_VERIFY_RETURN_VALUES=no
26  coverage_setting=''
27
28  while getopts ":c:n:s:CDERS" options; do
29    case "${options}" in
30      c)
31        PBT_STATEFUL_STEP_COUNT=${OPTARG}
32        ;;
33      n)
34        PBT_MAX_EXAMPLES=${OPTARG}
35        ;;
36      s)
37        PBT_SEED=${OPTARG}
38        ;;
39      C)
40        coverage_setting="--coverage"
41        ;;
42      D)
43        PBT_DEBUG=yes
44        ;;
45      E)
46        PBT_VERIFY_EVENTS=yes
47        ;;
48      R)
49        PBT_VERIFY_RETURN_VALUES=yes
50        ;;
51      S)
52        PBT_PROFILE="shrinking"
53        ;;
54      :)
55        echo "Error: -${OPTARG} requires an argument."
56        usage
57        exit 1
58        ;;
59
60      *)
61        echo Invalid arguments!
62        usage
63        exit 1
64        ;;
65    esac
66  done
67
68  shift $(expr $OPTIND - 1 )
69
70  if [ "$#" -eq 0 ]; then
71    echo No tests specified for execution!
72    usage
73    exit 1
74  fi
75  echo --- Environment
76  export PBT_SEED PBT_MAX_EXAMPLES PBT_STATEFUL_STEP_COUNT PBT_PROFILE PBT_VERIFY_EVENTS PBT_VERIFY_RETURN_VALUES
        PBT_DEBUG
77  env | grep ^PBT
```

```bash
78
79  extra_args=''
80
81  if [ "$PBT_DEBUG" == "yes" ]; then
82    extra_args="-s"
83  fi
84
85  echo —— Running tests
86  pytest --hypothesis-profile=$PBT_PROFILE $* $extra_args $coverage_setting
```

Appendix B

Bug analysis

B.1 BitAseanToken

This contract counts with two standard deviations. The first is a missing firing event Approval for a valid approval, detected by all testing agents. The second is related with a missing return value for a valid transfer, which was detected only by Brownie.

B.1.1 Property-based testing

Bug ID	Rule	Bug type	Info
BAS1	approve	Absent event	Approval event not fired for valid approval
BAS2	transfer	Absent return value	Fails to return a value for valid transfer

B.1.1.1 BAS1

The assertion error and the correspondent falsifying example are illustrated in Listing B.2 and B.3 for bug id BAS1. In Listing B.1 is possible to see that contract's implementation of approve() does not fire Approval event. Thus, this implementation is not compliant with the ERC−20 standard.

Listing B.1: BitAseanToken approve() implementation

```solidity
function approve(address _spender, uint256 _value)
  returns (bool success) {
  allowance[msg.sender][_spender] = _value;
  return true;
}
```

73

Listing B.2: BAS1 - Assertion error

```
Traceback (most recent call last):
  File "/home/celioggr/erc20-pbt/erc20_pbt.py", line 114, in
      rule_approveAndTransferAll
    self.rule_approve(st_owner, st_spender, amount)
  File "/home/celioggr/erc20-pbt/erc20_pbt.py", line 95, in rule_approve
    if self.DEBUG:
  File "/home/celioggr/.local/lib/python3.8/site-packages/hypothesis/stateful.
      py", line 594, in rule_wrapper
    return f(*args, **kwargs)
  File "/home/celioggr/erc20-pbt/erc20_pbt.py", line 102, in rule_approve
    self.verifyEvent(
AssertionError: Approval: event was not fired
```

Listing B.3: BAS1 - Falsifying example

```
Falsifying example:
state = BrownieStateMachine()
state.rule_approveAndTransferAll(st_owner=<Account '0
    x66aB6D9362d4F35596279692F0251Db635165871'>, st_receiver=<Account '0
    x66aB6D9362d4F35596279692F0251Db635165871'>, st_spender=<Account '0
    x0063046686E46Dc6F15918b61AE2B121458534a5'>)
state.teardown()
```

B.1.1.2 BAS2

Bug ID **BAS2**, is related with a return value that should have been sent. In Listing B.4, **transfer ()** interface signature explicitly says that upon success the transaction should return true.

Listing B.4: ERC-20 transfer signature

```
function transfer(address _to, uint256 _value) public returns (bool success)
```

The token contract fails to return such Boolean value, thus incurring in a violation that is detected when running the test with the option −R, which enables the verification of returned values upon function calls. Falsifying examples and assertion errors are detailed in Listing B.5 and B.6.

Listing B.5: BAS2 - Falsifying example

```
Falsifying example:
state = BrownieStateMachine()
state.rule_transferAll(st_receiver=<Account '0
    x66aB6D9362d4F35596279692F0251Db635165871'>, st_sender=<Account '0
    x66aB6D9362d4F35596279692F0251Db635165871'>)
state.teardown()
```

Listing B.6: BAS2 - Assertion Error

```
Traceback (most recent call last):
  File "/home/celioggr/erc20-pbt/erc20_pbt.py", line 110, in rule_transferAll
    self.rule_transfer(st_sender, st_receiver, self.balances[st_sender])
  File "/home/celioggr/erc20-pbt/erc20_pbt.py", line 41, in rule_transfer
    if self.DEBUG:
  File "/home/celioggr/erc20-pbt/erc20_pbt.py", line 56, in rule_transfer
    self.verifyReturnValue(tx, True)
AssertionError: return value : expected value True, actual value was None
```

B.1.2 OpenZeppelin unit testing

Bug ID	Function	Bug type	Info	test ID
OZ_BAS1	approve	Absent event	Approval event not fired for valid approval	1,2

B.1.2.1 OZ_BAS1

The test IDs for bug ID **OZ_BAS1** are detailed in Listings B.7 and B.8. This bug is with bug ID **BAS1** (B.1.1.1).

Listing B.7: OZ_BAS1 - test ID 1

```
1) Contract: BitAseanToken
      approve
        when the spender is not the zero address
          when the sender has enough balance
            emits an approval event:

      No 'Approval' events found
      + expected - actual
```

Listing B.8: OZ_BAS1 - test ID 2

```
2) Contract: BitAseanToken
     approve
       when the spender is not the zero address
         when the sender does not have enough balance
           emits an approval event:

   No 'Approval' events found
   + expected - actual
```

B.1.3 Consensys unit testing

Bug ID	Function	Assertion failure	Info	test ID
CS_BAS1	approve	Absent event	Approval event not fired for valid approval	1

B.1.3.1 CS_BAS1

The test ID for bug ID CS_BAS1 is detailed in Listing B.9. This bug is with bug ID BAS1 (B.1.1.1).

Listing B.9: CS_BAS1 - test ID 1

```
1) Contract: BitAseanToken
     events: should fire Approval event properly:
   TypeError: Cannot read property 'args' of undefined
```

B.2 BNB

This contract reports 5 bugs. PBT finds 3 bugs, all related to the fact that an approval or transfer of 0 tokens reverts in approve, transfer, and transferFrom. The last two, are related with a missing return value for a valid transfer and an absent Approval event for a valid approve. The bug in transferFrom is not exposed by Consensys or OpenZeppelin, and the bug in approve is not exposed by OpenZeppelin.

Bug ID	Rule	Assertion failure	Info
BNB1	approve	Absent event	missing Approval for valid transaction
BNB2	approve	Operation not allowed	reverts approval of 0 tokens
BNB3	transfer	Absent return value	fails to return value for valid transfer
BNB4	transfer	Operation not allowed	reverts transfer of 0 tokens
BNB5	transferFrom	Operation not allowed	reverts transfer of 0 tokens

B.2.1.1   BNB1

This bug is with a missing Approval event for a valid approval (Listing B.11 line 4). In Listing B.10 it is possible to see that approve() does not emit Approval event.

Listing B.10: BNB approve() source code

```
1    function approve(address _spender, uint256 _value)
2       returns (bool success) {
3   if (_value <= 0) throw;
4       allowance[msg.sender][_spender] = _value;
5       return true;
6   }
```

The assertion error and the correspondent falsifying example are illustrated in Listing B.11 in Listing B.12 for bug ID BNB1.

Listing B.11: BNB1 - Assertion error

```
Traceback (most recent call last):
  File "/home/celioggr/erc20-pbt/erc20_pbt.py", line 114, in
      rule_approveAndTransferAll
    self.rule_approve(st_owner, st_spender, amount)
  File "/home/celioggr/erc20-pbt/erc20_pbt.py", line 95, in rule_approve
    if self.DEBUG:
  File "/home/celioggr/.local/lib/python3.8/site-packages/hypothesis/stateful.
      py", line 594, in rule_wrapper
    return f(*args, **kwargs)
  File "/home/celioggr/erc20-pbt/erc20_pbt.py", line 102, in rule_approve
    self.verifyEvent(
AssertionError: Approval: event was not fired
```

Listing B.12: BNB1 - Falsifying example

```
Falsifying example:
state = BrownieStateMachine()
state.rule_approveAndTransferAll(st_owner=<Account '0
    x66aB6D9362d4F35596279692F0251Db635165871'>, st_receiver=<Account '0
    x66aB6D9362d4F35596279692F0251Db635165871'>, st_spender=<Account '0
    x0063046686E46Dc6F15918b61AE2B121458534a5'>)
state.teardown()
```

B.2.1.2   BNB2

This bug is with a valid approval of zero tokens that is reverted. In Listing B.10 line 3, is possible to see that if _value $<=$ 0, then the current call will throw. The assertion error and the correspondent falsifying example are illustrated in Listing B.13 in Listing B.14 for bug ID BNB2. In particular, the falsifying example in Listing B.14 uses rule_approveAndTransferAll, which tries to approve all tokens from the involved account. The amount of tokens at that stage is zero since all tokens are assigned to accounts[0].

Listing B.13: BNB2 - Assertion error

```
Traceback (most recent call last):
  File "/home/celioggr/erc20-pbt/erc20_pbt.py", line 114, in
      rule_approveAndTransferAll
    self.rule_approve(st_owner, st_spender, amount)
  File "/home/celioggr/erc20-pbt/erc20_pbt.py", line 95, in rule_approve
    if self.DEBUG:
  File "/home/celioggr/.local/lib/python3.8/site-packages/hypothesis/stateful.
      py", line 594, in rule_wrapper
    return f(*args, **kwargs)
  File "/home/celioggr/erc20-pbt/erc20_pbt.py", line 98, in rule_approve
    tx = self.contract.approve(
brownie.exceptions.VirtualMachineError: revert
```

Listing B.14: BNB2 - Falsifying example

```
Falsifying example:
state = BrownieStateMachine()
state.rule_approveAndTransferAll(st_owner=<Account '0
    x33A4622B82D4c04a53e170c638B944ce27cffce3'>, st_receiver=<Account '0
    x66aB6D9362d4F35596279692F0251Db635165871'>, st_spender=<Account '0
    x0063046686E46Dc6F15918b61AE2B121458534a5'>)
state.teardown()
```

B.2.1.3   BNB3

This bug is with a missing return boolean value for a valid transfer. In Listing B.15 it is possible
to see that **transfer** () does not return true when a valid transfer takes place. The assertion error
and the correspondent falsifying example are illustrated in Listing B.16 and in Listing B.17 for
bug ID BNB3.

Listing B.15: BNB3 - transfer() source code

```
1    function transfer(address _to, uint256 _value) {
2      if (_to == 0x0) throw;
3  if (_value <= 0) throw;
4      if (balanceOf[msg.sender] < _value) throw;
5      if (balanceOf[_to] + _value < balanceOf[_to]) throw;
6      balanceOf[msg.sender] = SafeMath.safeSub(balanceOf[msg.sender], _value);
7      balanceOf[_to] = SafeMath.safeAdd(balanceOf[_to], _value);
8      Transfer(msg.sender, _to, _value);
9    }
```

Listing B.16: BNB3 - Assertion error

```
Traceback (most recent call last):
  File "/home/celioggr/erc20-pbt/erc20_pbt.py", line 110, in rule_transferAll
    self.rule_transfer(st_sender, st_receiver, self.balances[st_sender])
  File "/home/celioggr/erc20-pbt/erc20_pbt.py", line 41, in rule_transfer
    if self.DEBUG:
  File "/home/celioggr/.local/lib/python3.8/site-packages/hypothesis/stateful.
    py", line 594, in rule_wrapper
    return f(*args, **kwargs)
  File "/home/celioggr/erc20-pbt/erc20_pbt.py", line 56, in rule_transfer
    self.verifyReturnValue(tx, True)
AssertionError: return value : expected value True, actual value was None
```

Listing B.17: BNB3 - Falsifying example

```
Falsifying example:
state = BrownieStateMachine()
state.rule_transferAll(st_receiver=<Account '0
    x66aB6D9362d4F35596279692F0251Db635165871'>, st_sender=<Account '0
    x66aB6D9362d4F35596279692F0251Db635165871'>)
state.teardown()
```

B.2.1.4 BNB4

This bug is with a reverted transaction for a valid transfer of zero tokens (when using transfer ()).
In Listing B.15 line 3 it is possible to see that if _value <= 0 the function will throw. The
assertion error and the correspondent falsifying example are illustrated in Listing B.18 and in
Listing B.19 for bug ID BNB4.

Listing B.18: BNB4 - Assertion error

```
Traceback (most recent call last):
  File "/home/celioggr/erc20-pbt/erc20_pbt.py", line 110, in rule_transferAll
    self.rule_transfer(st_sender, st_receiver, self.balances[st_sender])
  File "/home/celioggr/erc20-pbt/erc20_pbt.py", line 41, in rule_transfer
    if self.DEBUG:
  File "/home/celioggr/.local/lib/python3.8/site-packages/hypothesis/stateful.
     py", line 594, in rule_wrapper
    return f(*args, **kwargs)
  File "/home/celioggr/erc20-pbt/erc20_pbt.py", line 47, in rule_transfer
    tx = self.contract.transfer(
brownie.exceptions.VirtualMachineError: revert
Trace step -1, program counter 2061:
  File "contracts/BNB.sol", line 83, in BNB.transfer:
```

Listing B.19: BNB4 - Falsifying example

```
Falsifying example:
state = BrownieStateMachine()
state.rule_transferAll(st_receiver=<Account '0
   x66aB6D9362d4F35596279692F0251Db635165871'>, st_sender=<Account '0
   x33A4622B82D4c04a53e170c638B944ce27cffce3'>)
state.teardown()
```

B.2.1.5 BNB5

This bug is with a reverted transaction for a valid transfer of zero tokens (when using transfer-
From()). In Listing B.15 line 3 it is possible to see that if _value <= 0 the function will throw.
The assertion error and the correspondent falsifying example are illustrated in Listing B.20 and
in Listing B.21 for bug ID BNB4.

Listing B.20: BNB5 - Assertion error

```
Falsifying example:
state = BrownieStateMachine()
state.rule_transferFrom(st_amount=0, st_owner=<Account '0
    x66aB6D9362d4F35596279692F0251Db635165871'>, st_receiver=<Account '0
    x66aB6D9362d4F35596279692F0251Db635165871'>, st_spender=<Account '0
    x66aB6D9362d4F35596279692F0251Db635165871'>)
state.teardown()
```

Listing B.21: BNB5 - Falsifying example

```
Traceback (most recent call last):
  File "/home/celioggr/erc20-pbt/erc20_pbt.py", line 76, in rule_transferFrom
    tx = self.contract.transferFrom(
brownie.exceptions.VirtualMachineError: revert
Trace step -1, program counter 1088:
  File "contracts/BNB.sol", line 103, in BNB.transferFrom:
      /* A contract attempts to get the coins */
      function transferFrom(address _from, address _to, uint256 _value)
          returns (bool success) {
          if (_to == 0x0) throw;
    if (_value <= 0) throw;
          if (balanceOf[_from] < _value) throw;
          if (balanceOf[_to] + _value < balanceOf[_to]) throw;
          if (_value > allowance[_from][msg.sender]) throw;
```

B.2.2   OpenZeppelin unit testing

Bug ID	Function	Assertion failure	Info	test ID
OZ_BNB1	transfer	Operation not allowed	revert for valid transaction	1,2
OZ_BNB2	approve	Absent event	event not fired for valid transaction	3,4

B.2.2.1   OZ_BNB1

This bug is with bug ID **BNB4** (B.2.1.4). The test IDs with bug ID **OZ_BNB1** are detailed in
Listings B.22, B.23.

Listing B.22: OZ_BNB1 - test ID 1

```
1) Contract: BNB
    transfer
      when the recipient is not the zero address
        when the sender transfers zero tokens
          transfers the requested amount:
  Error: Returned error: VM Exception while processing transaction: revert
```

Listing B.23: OZ_BNB1 - test ID 2

```
2) Contract: BNB
      transfer
        when the recipient is not the zero address
          when the sender transfers zero tokens
            emits a transfer event:
    Error: Returned error: VM Exception while processing transaction: revert
```

B.2.2.2 OZ_BNB2

This bug is with bug ID BNB1(B.2.1.1). The test IDs with bug ID OZ_BNB2 are detailed in
Listings B.24, B.25.

Listing B.24: OZ_BNB2 - test ID 3

```
3) Contract: BNB
      approve
        when the spender is not the zero address
          when the sender has enough balance
            emits an approval event:
      No 'Approval' events found
      + expected - actual
      -false
      +true
```

Listing B.25: OZ_BNB2 - test ID 4

```
4) Contract: BNB
     approve
       when the spender is not the zero address
         when the sender does not have enough balance
           emits an approval event:
   No 'Approval' events found
   + expected - actual
   -false
   +true
```

B.2.3   Consensys unit testing

Bug ID	Function	Assertion failure	Info	test ID
CS_BNB1	transfer	Operation not allowed	revert for valid transaction	1,3
CS_BNB2	approve	Operation not allowed	revert for valid transaction	2
CS_BNB3	approve	Absent event	event not fired for valid transaction	4

B.2.3.1   CS_BNB1

This bug is with bug ID BNB4 (B.2.1.4). The test IDs with bug ID OZ_BNB1 are detailed in Listings B.26, B.27.

Listing B.26: CS_BNB1 - test ID 1

```
1) Contract: BNB
      transfers: should handle zero-transfers normally:
   Transaction: 0
      x0fc6e7688aa97d38e810d67ef91c93ad87abe32f9791b2a44ac34d8a8fba0b35
      exited with an error (status 0).
   Please check that the transaction:
   - satisfies all conditions set by Solidity 'require' statements.
   - does not trigger a Solidity 'revert' statement.
```

Listing B.27: CS_BNB1 - test ID 3

```
3) Contract: BNB
     events: should fire Transfer event normally on a zero transfer:
   Transaction: 0
       xabbd340f9d8948fa576852aadfdd09b792245e384b042c1d9a00254f5e3c10ea exited
       with an error (status 0).
   Please check that the transaction:
   - satisfies all conditions set by Solidity 'require' statements.
   - does not trigger a Solidity 'revert' statement.
```

B.2.3.2   CS_BNB2

This bug is with bug ID BNB2(B.2.3.2). The test ID with bug ID CS_BNB2 is detailed in Listing B.28.

Listing B.28: OZ_BNB2 - test ID 2

```
2) Contract: BNB
     approvals: allow accounts[1] 100 to withdraw from accounts[0]. Withdraw 60
         and then approve 0 & attempt transfer.:
   Transaction: 0
       x7852d6a15712ba50b4c397721746c057c1f6cbcb865cf2c673feb27dcbc70969 exited
       with an error (status 0).
   Please check that the transaction:
   - satisfies all conditions set by Solidity 'require' statements.
   - does not trigger a Solidity 'revert' statement.
```

B.2.3.3   CS_BNB3

This bug is with bug ID BNB1 (B.2.1.1). The test ID with bug ID CS_BNB3 is detailed in Listing B.29.

Listing B.29: OZ_BNB3 - test ID 4

```
4) Contract: BNB
     events: should fire Approval event properly:
   TypeError: Cannot read property 'args' of undefined
```

B.3 FuturXe

This contract is particular interesting because of its poorly written code and lack of compliance with the ERC-20 standard. From valid transactions not taking place, to absent reverts and buggy implementations, the contract counts with a total of 4 bugs. All testing agents reported the same bugs.

B.3.1 Property-based testing

Bug ID	Rule	Assertion failure	Info
FXE1	transfer	Absent revert	returns false instead of reverting
FXE2	transferFrom	Invalid Operation allowed	corrupts balances and allowances
FXE3	transfer	Operation not allowed	does not allow valid transfer
FXE4	transferFrom	Absent revert	returns false instead of reverting

B.3.1.1 FXE1

According to ERC-20 (Listing B.30), when the caller's address has not enough tokens to transfer the transaction should revert.

Listing B.30: EIP20:ERC20 - transfer()

```
The function SHOULD throw if the message caller's account balance does not
    have enough tokens to spend.
```

In Listing B.31 we see that conditional statement in line 3 returns false instead of reverting the current call with throw or require statements.

Listing B.31: FuturXe - transfer() source code

```
1   function transfer(address to, uint value) returns (bool success) {
2       if (frozenAccount[msg.sender]) return false;
3       if(balances[msg.sender] < value) return false;
4       if(balances[to] + value < balances[to]) return false;
5       balances[msg.sender] -= value;
6       balances[to] += value;
7       Transfer(msg.sender, to, value);
8       return true;
9   }
```

The assertion error and the correspondent falsifying example are illustrated in Listing B.32

and Listing B.33 for bug ID **FXE1**.

Listing B.32: FXE1 - Assertion error

```
raise AssertionError("Transaction did not revert") from None
AssertionError: Transaction did not revert
```

Listing B.33: FXE1 - Falsifying example

```
state = BrownieStateMachine()
transfer(0x33A4622B82D4c04a53e170c638B944ce27cffce3, 0
    x66aB6D9362d4F35596279692F0251Db635165871, 1)
state.rule_transfer(st_amount=1, st_receiver=<Account '0
    x66aB6D9362d4F35596279692F0251Db635165871'>, st_sender=<Account '0
    x33A4622B82D4c04a53e170c638B944ce27cffce3'>)
```

B.3.1.2   FXE2

The assertion error and the correspondent falsifying example are illustrated in Listing B.35 and B.36 for bug ID **FXE2**.

From the falsifying example in Listing B.36 we see that Hypothesis tried to transfer 231 tokens without having a valid allowance. In Listing B.34 we see that there is a check for the allowance in question in line 5. However, the allowance is set to 0 which is not greater than or equal to 231, thus the current call is allowed to continue execution and withdraw tokens from the owner account. This invalid conditional statement might corrupt account balances and allowances. Hence, the conditional statement should be allowed[from][msg.sender] < value instead of allowed[from][msg.sender] >= value.

Listing B.34: FXE2 - transferFrom() with bug conditional statement

```
1   function transferFrom(address from, address to, uint value)
2     returns (bool success) {
3     if (frozenAccount[msg.sender]) return false;
4     if(balances[from] < value) return false;
5     if( allowed[from][msg.sender] >= value ) return false;
6     if(balances[to] + value < balances[to]) return false;
7     balances[from] -= value;
8     allowed[from][msg.sender] -= value;
9     balances[to] += value;
10    Transfer(from, to, value);
11    return true;
12  }
```

Listing B.35: FXE2 - Assertion error

```
AssertionError("{} : expected value {}, actual value was {}".format(msg,
    expected, actual))
AssertionError: balanceOf(0x66aB6D9362d4F35596279692F0251Db635165871) :
    expected value 1000, actual value was 769
```

Listing B.36: FXE2 - Falsifying example

```
state = BrownieStateMachine()
state.rule_transferFrom(st_amount=231, st_owner=<Account '0
    x66aB6D9362d4F35596279692F0251Db635165871'>, st_receiver=<Account '0
    xA868bC7c1AF08B8831795FAC946025557369F69C'>, st_spender=<Account '0
    x844ec86426F076647A5362706a04570A5965473B'>)
state.teardown()
```

B.3.1.3   FXE3

The assertion error and the correspondent falsifying example are illustrated in Listing B.38 and B.39 for bug ID FX3.

In Listing B.37 line 5 it is possible to see that if allowed[from][msg.sender] is greater than or equal to value, then the function returns false. Meaning that, it is not possible to transfer all the tokens for a valid allowance.

Listing B.37: FXE3 - transferFrom() source code

```
1   function transferFrom(address from, address to, uint value)
2   returns (bool success) {
3     if (frozenAccount[msg.sender]) return false;
4     if(balances[from] < value) return false;
5     if(allowed[from][msg.sender] >= value ) return false;
6     if(balances[to] + value < balances[to]) return false;
7     balances[from] -= value;
8     allowed[from][msg.sender] -= value;
9     balances[to] += value;
10    Transfer(from, to, value);
11    return true;
12  }
```

Listing B.38: FXE3 - Assertion error

```
Traceback (most recent call last):
  File "/home/celioggr/erc20-pbt/erc20_pbt.py", line 115, in
     rule_approveAndTransferAll
    self.rule_transferFrom(st_spender, st_owner, st_receiver, amount)
  File "/home/celioggr/erc20-pbt/erc20_pbt.py", line 64, in rule_transferFrom
    if self.DEBUG:
  File "/home/celioggr/.local/lib/python3.8/site-packages/hypothesis/stateful.
     py", line 594, in rule_wrapper
    return f(*args, **kwargs)
  File "/home/celioggr/erc20-pbt/erc20_pbt.py", line 81, in rule_transferFrom
    self.verifyAllowance(st_owner, st_spender, -st_amount)
AssertionError: allowance(0x66aB6D9362d4F35596279692F0251Db635165871,0
    x33A4622B82D4c04a53e170c638B944ce27cffce3) : expected value 0, actual
    value was 1000
```

Listing B.39: FXE3 - Falsifying example

```
Falsifying example:
state = BrownieStateMachine()
state.rule_approveAndTransferAll(st_owner=<Account '0
    x66aB6D9362d4F35596279692F0251Db635165871'>, st_receiver=<Account '0
    x66aB6D9362d4F35596279692F0251Db635165871'>, st_spender=<Account '0
    x33A4622B82D4c04a53e170c638B944ce27cffce3'>)
state.teardown()
```

B.3.1.4 FXE4

Same as bug ID **FXE1** (B.3.1.1), but using transferFrom() The assertion error and the correspondent falsifying example are illustrated in Listing B.40 and B.41 for bug ID **FXE4**.

Listing B.40: FXE4 - Assertion error

```
Traceback (most recent call last):
  File "/home/celioggr/.local/lib/python3.8/site-packages/hypothesis/stateful.
     py", line 165, in run_state_machine
    result = rule.function(machine, **data)
  File "/home/celioggr/erc20-pbt/erc20_pbt.py", line 90, in rule_transferFrom
    self.contract.transferFrom(
  File "/home/celioggr/erc20-pbt/erc20_pbt.py", line 350, in alt_exit
    return f(self, exc_type, exc_value, traceback)
  File "brownie/test/managers/runner.py", line 64, in __exit__
    raise AssertionError("Transaction did not revert")
AssertionError: Transaction did not revert
```

Listing B.41: FXE4 - Falsifying example

```
Falsifying example:
state = BrownieStateMachine()
state.rule_transferFrom(st_amount=7, st_owner=<Account '0
    x66aB6D9362d4F35596279692F0251Db635165871'>, st_receiver=<Account '0
    x0063046686E46Dc6F15918b61AE2B121458534a5'>, st_spender=<Account '0
    x66aB6D9362d4F35596279692F0251Db635165871'>)
state.teardown()
```

B.3.2 OpenZeppelin unit testing

Bug ID	Function	Assertion failure	Info	test ID
OZ_FXE1	transfer/transferFrom	Absent revert	does not revert	1,7,8,9
OZ_FXE2	transferFrom	Invalid Operation	allows invalid operation	4,5
OZ_FXE3	transfer	Operation not allowed	does not allow valid transfer	2,3,6

B.3.2.1 OZ_FXE1

The test IDs for bug ID **OZ_FXE1** are detailed in Listings B.42, B.43, B.44 and B.45. This bug is with bug ID **FXE1** (B.3.1.1) and **FXE4** (B.3.1.4).

Listing B.42: OZ_FXE1 - test ID 1

```
1) Contract: FuturXe
       transfer
         when the recipient is not the zero address
          when the sender does not have enough balance
            reverts:
       AssertionError: Expected an exception but none was received
```

Listing B.43: OZ_FXE1 - test ID 7

```
7) Contract: FuturXe
       transfer from
          when the token owner is not the zero address
           when the recipient is not the zero address
            when the spender has enough approved balance
              when the token owner does not have enough balance
               reverts:
       AssertionError: Expected an exception but none was received
```

Listing B.44: OZ_FXE1 - test ID 8

```
8) Contract: FuturXe
      transfer from
          when the token owner is not the zero address
            when the recipient is not the zero address
              when the spender does not have enough approved balance
                when the token owner has enough balance
                  reverts:
      AssertionError: Expected an exception but none was received
```

Listing B.45: OZ_FXE1 - test ID 9

```
9) Contract: FuturXe
      transfer from
          when the token owner is not the zero address
            when the recipient is not the zero address
              when the spender does not have enough approved balance
                when the token owner does not have enough balance
                  reverts:
      AssertionError: Expected an exception but none was received
```

B.3.2.2 OZ_FXE2

The test IDs for bug ID **OZ_FXE2** are detailed in Listings B.46 and B.47. This bug is related
with bug ID **FXE2** (B.3.1.2) where invalid operations are allowed resulting in account balances
and allowances being corrupted.

Listing B.46: OZ_FXE2 - test ID 4

```
4) Contract: FuturXe
      transfer from
          when the token owner is not the zero address
            when the recipient is not the zero address
              when the spender has enough approved balance
                when the token owner has enough balance
                  transfers the requested amount:
      AssertionError: expected '100' to equal '0'
      + expected - actual
      -100
      +0
```

Listing B.47: OZ_FXE2 - test ID 5

```
5) Contract: FuturXe
      transfer from
        when the token owner is not the zero address
          when the recipient is not the zero address
            when the spender has enough approved balance
              when the token owner has enough balance
                decreases the spender allowance:
      AssertionError: expected '100' to equal '0'
      + expected - actual
      -100
      +0
```

B.3.2.3   OZ_FXE3

The test IDs for bug ID **OZ_FXE3** are detailed in Listings B.48, B.49 and B.50. This bug is related with bug ID **FXE3** (B.3.1.3) where valid operations are not allowed.

Listing B.48: OZ_FXE3 - test ID 2

```
2) Contract: FuturXe
    transfer
      when the recipient is not the zero address
        when the sender transfers all balance
          emits a transfer event:
    Event argument 'from' not found
    + expected - actual
    -false
    +true
```

Listing B.49: OZ_FXE3 - test ID 3

```
3) Contract: FuturXe
    transfer
      when the recipient is not the zero address
        when the sender transfers zero tokens
          emits a transfer event:
    Event argument 'from' not found
    + expected - actual
    -false
    +true
```

Listing B.50: OZ_FXE3 - test ID 6

```
6) Contract: FuturXe
     transfer from
       when the token owner is not the zero address
         when the recipient is not the zero address
           when the spender has enough approved balance
             when the token owner has enough balance
               emits a transfer event:
     No 'Transfer' events found
     + expected − actual
     −false
     +true
```

B.3.3 Consensys unit testing

Bug ID	Function	Assertion failure	Info	test ID
CS_FXE1	transfer/transferFrom	Absent revert	does not revert	1,5,6
CS_FXE2	transferFrom	Assertion error balances	buggy transferFrom()	2,3,4,7

B.3.3.1 CS_FXE1

The test IDs for bug ID **CS_FXE1** are detailed in Listings B.51, B.52 and B.53. This bug is related with bug ID **FXE1** (B.3.1.1) and **FXE4** (B.3.1.4) where invalid transactions return **false** instead of reverting.

Listing B.51: CS_FXE1 - test ID 1

```
1) Contract: FuturXe
     transfers: should fail when trying to transfer 10001 to accounts[1] with
         accounts[0] having 10000:
   AssertionError: Expected revert not received
```

Listing B.52: CS_FXE1 - test ID 5

```
5) Contract: FuturXe
       approvals: attempt withdrawal from account with no allowance (should
           fail):
     AssertionError: Expected revert not received
```

Listing B.53: CS_FXE1 - test ID 6

```
6) Contract: FuturXe
     approvals: allow accounts[1] 100 to withdraw from accounts[0]. Withdraw 60
         and then approve 0 & attempt transfer.:
   AssertionError: Expected revert not received
```

B.3.3.2 CS_FXE2

The test ID3 for bug ID **CS_FXE2** are detailed in Listings B.54,B.55 and B.56 and B.57. This
bug is related with bug ID **FXE2** (B.5.1.2) where invalid transfers are allowed.

Listing B.54: CS_FXE2 - test ID 2

```
2) Contract: FuturXe
       approvals: msg.sender approves accounts[1] of 100 & withdraws 20 once.:
       AssertionError: expected 100 to equal 80
       + expected - actual
       -100
       +80
```

Listing B.55: CS_FXE2 - test ID 3

```
3) Contract: FuturXe
       approvals: msg.sender approves accounts[1] of 100 & withdraws 20 twice
         .:
       AssertionError: expected 100 to equal 80
       + expected - actual
       -100
       +80
```

Listing B.56: CS_FXE2 - test ID 4

```
4) Contract: FuturXe
       approvals: msg.sender approves accounts[1] of 100 & withdraws 50 & 60
         (2nd tx should fail):
       AssertionError: expected 100 to equal 50
       + expected - actual
       -100
       +50
```

Listing B.57: CS_FXE2 - test ID 7

```
7) Contract: FuturXe
      approvals: msg.sender approves accounts[1] of max (2^256 − 1) &
          withdraws 20:
      AssertionError: expected 0 to equal 20
      + expected − actual
      −0
      +20
```

B.4 HBToken

HuobiToken has two bugs, reported by all testing agents. These are related with transactions that should **revert** when invalid conditions are met. Instead, the contract's implementation returns a Boolean value to handle such scenarios.

B.4.1 Property-based testing

Bug ID	Rule	Assertion failure	Info
HT1	transfer	Absent revert	Returned false instead of reverting
HT2	transferFrom	Absent revert	Returned false instead of reverting

B.4.1.1 HT1

From the **transfer** () source code in Listing B.58, we can see that when invalid inputs exist (p.e not enough balance) the function returns **false** instead of reverting (line 8). The assertion error and the correspondent falsifying example are illustrated in Listing B.59 and the correspondent falsifying example in Listing B.60 for bug ID **HT1**.

Listing B.58: HuobiToken transfer() source code

```
function transfer(address _to, uint _value) returns (bool) {
  //Default assumes totalSupply can't be over max (2^256 − 1).
  if (balances[msg.sender] >= _value && balances[_to] + _value >= balances[
      _to]) {
    balances[msg.sender] −= _value;
    balances[_to] += _value;
    Transfer(msg.sender, _to, _value);
    return true;
  } else { return false; }
}
```

Listing B.59: HT1 - Assertion error

```
self = <brownie.test.managers.runner.RevertContextManager object at 0
    x7fea0a2120d0>, exc_type = None, exc_value = None
traceback = None


    def alt_exit(self, exc_type, exc_value, traceback):
        if exc_type is VirtualMachineError:
            exc_value.__traceback__.tb_next = None
            if exc_value.revert_type != "revert":
                return False
>           return f(self, exc_type, exc_value, traceback)
E           AssertionError: Transaction did not revert
```

Listing B.60: HT1 - Falsifying example

```
Falsifying example:
state = BrownieStateMachine()
state.rule_transfer(st_amount=1, st_receiver=<Account '0
    x23BB2Bb6c340D4C91cAa478EdF6593fC5c4a6d4B'>, st_sender=<Account '0
    x33A4622B82D4c04a53e170c638B944ce27cffce3'>)
state.teardown()
```

B.4.1.2 HT2

Same as HT1 (B.4.1.1) but for transferFrom(). The assertion error and the correspondent falsifying example are illustrated in Listing B.61 and B.62 for bug ID HT2.

Listing B.61: HT2 - Assertion error

```
self = <brownie.test.managers.runner.RevertContextManager object at 0
    x7f38e45e2760>, exc_type = None, exc_value = None
traceback = None


    def alt_exit(self, exc_type, exc_value, traceback):
        if exc_type is VirtualMachineError:
            exc_value.__traceback__.tb_next = None
            if exc_value.revert_type != "revert":
                return False
>           return f(self, exc_type, exc_value, traceback)
E           AssertionError: Transaction did not revert
```

Listing B.62: HT2 - Falsifying example

```
Falsifying example:
state = BrownieStateMachine()
state.rule_transferFrom(st_amount=1, st_owner=<Account '0
    x33A4622B82D4c04a53e170c638B944ce27cffce3'>, st_receiver=<Account '0
    x33A4622B82D4c04a53e170c638B944ce27cffce3'>, st_spender=<Account '0
    x66aB6D9362d4F35596279692F0251Db635165871'>)
state.teardown()
```

B.4.2 OpenZeppelin unit testing

Bug ID	Function	Assertion failure	Info	test ID
OZ_HT1	transfer/transferFrom	Absent revert	did not revert	1,2,3,4

B.4.2.1 OZ_HT1

The test IDs concerning bug ID **OZ_HT1** are related with bug ID **HT1** (B.4.1.1) and are detailed in Listings B.63, B.64, B.65 and B.66.

Listing B.63: OZ_HT1 - test ID 1

```
1) Contract: HBToken
   transfer
     when the recipient is not the zero address
       when the sender does not have enough balance
         reverts:
 AssertionError: Expected an exception but none was received
```

Listing B.64: OZ_HT1 - test ID 2

```
2) Contract: HBToken
     transfer from
         when the token owner is not the zero address
         when the recipient is not the zero address
           when the spender has enough approved balance
             when the token owner does not have enough balance
               reverts:
 AssertionError: Expected an exception but none was received
```

Listing B.65: OZ_HT1 - test ID 3

```
3) Contract: HBToken
   transfer from
     when the token owner is not the zero address
       when the recipient is not the zero address
         when the spender does not have enough approved balance
           when the token owner has enough balance
             reverts:
  AssertionError: Expected an exception but none was received
```

Listing B.66: OZ_HT1 - test ID 4

```
4) Contract: HBToken
   transfer from
     when the token owner is not the zero address
       when the recipient is not the zero address
         when the spender does not have enough approved balance
           when the token owner does not have enough balance
             reverts:
  AssertionError: Expected an exception but none was received
```

B.4.3 Consensys unit testing

Bug ID	Function	Assertion failure	Info	test ID
CS_HT1	transfer/transferFrom	Absent revert	did not revert	1,2,3,4

B.4.3.1 CS_HT1

The test IDs concerning bug ID CS_HT1 are related with bug ID HT1 (B.4.1.1) detailed in Listings B.67, B.68, B.69 and B.70.

Listing B.67: CS_HT1 - test ID 1

```
1) Contract: HuobiToken
   transfers: should fail when trying to transfer 10001 to accounts[1] with
     accounts[0] having 10000:
  AssertionError: Expected revert not received
```

Listing B.68: CS_HT1 - test ID 2

```
2) Contract: HuobiToken
    approvals: msg.sender approves accounts[1] of 100 & withdraws 50 & 60 (2nd
        tx should fail):
  AssertionError: Expected revert not received
```

Listing B.69: CS_HT1 - test ID 3

```
3) Contract: HuobiToken
    approvals: attempt withdrawal from account with no allowance (should fail):
  AssertionError: Expected revert not received
```

Listing B.70: CS_HT1 - test ID 4

```
4) Contract: HuobiToken
    approvals: allow accounts[1] 100 to withdraw from accounts[0]. Withdraw 60
        and then approve 0 & attempt transfer.:
  AssertionError: Expected revert not received
```

B.5 InternetNodeToken

This contract fails to fire the **Approval** event when processing **approve()** transactions. Once again this function seems to reuse code from previous analyzed contracts. Also this token contracts fails to meet the ERC-20 standard when dealing with transfers of 0 tokens. Although the standard explicity states that these transfers should be treated as valid ones, the contract chooses to revert them. Consensys fails to detect the bug where **transferFrom()** reverts for a valid transfer when allowance(owner,msg.sender) = amount.

B.5.1 Property-based testing

Bug ID	Rule	Assertion failure	Info
INT1	transfer	Absent return value	Fails to return a value for valid transfer
INT2	approve	Absent event	Approval event not fired for valid approve
INT3	transfer	Operation not allowed	Reverts when balance(owner) = amount
INT4	transferFrom	Operation not allowed	Reverts when balance(owner) = amount
INT5	transferFrom	Operation not allowed	Reverts when allowances(owner) = amount

The stateful test reported an assertion error detailed in Listing B.72 and the correspondent falsifying example in Listing B.73 for bug ID INT1. In Listing B.71, it is possible to see that for valid transactions the function will never return true.

Listing B.71: InternetNodeToken - transfer() source code

```
1   function _transfer(address _from, address _to, uint _value) internal {
2     require (_to != 0x0);
3     require (balanceOf[_from] > _value);
4     require (balanceOf[_to] + _value > balanceOf[_to]);
5     require(!frozenAccount[_from]);
6     require(!frozenAccount[_to]);
7     balanceOf[_from] -= _value;
8     balanceOf[_to] += _value;
9     Transfer(_from, _to, _value);
10  }
```

Listing B.72: INT1 - Assertion error

```
Traceback (most recent call last):
  File "/home/celioggr/erc20-pbt/erc20_pbt.py", line 56, in rule_transfer
    self.verifyReturnValue(tx, True)
AssertionError: return value : expected value True, actual value was None
```

Listing B.73: INT1 - Falsifying example

```
Falsifying example:
state = BrownieStateMachine()
state.rule_transfer(st_amount=168, st_receiver=<Account '0
    x66aB6D9362d4F35596279692F0251Db635165871'>, st_sender=<Account '0
    x66aB6D9362d4F35596279692F0251Db635165871'>)
state.teardown()
```

B.5.1.2 INT2

Bug ID INT2, also fails to fire Approval event as in BAS1 (B.1.1.1) and SWFTC1 (B.7.1.1), due to code reuse of approve() function. The assertion error and the correspondent falsifying example is illustrated in Listing B.74 and B.75 for bug ID INT2.

Listing B.74: INT2 - Assertion error

```
    def verifyEvent(self, tx, eventName, data):
        if not eventName in tx.events:
>           raise AssertionError("{}: event was not fired".format(eventName))
E           AssertionError: Approval: event was not fired
```

Listing B.75: INT2 - Falsifying example

```
Falsifying example:
state = BrownieStateMachine()
state.rule_approve(st_amount=0, st_owner=<Account '0
    x66aB6D9362d4F35596279692F0251Db635165871'>, st_spender=<Account '0
    x66aB6D9362d4F35596279692F0251Db635165871'>)
state.teardown()
```

B.5.1.3 INT3

In Listing B.71 line 3, is possible to see that an account can't transfer all the tokens that holds. The resulting assertion error and falsifying example are detailed in Listing B.76 and B.77 respectively.

Listing B.76: INT3 - Assertion error

```
Traceback (most recent call last):
  File "/home/celioggr/erc20-pbt/erc20_pbt.py", line 110, in rule_transferAll
    self.rule_transfer(st_sender, st_receiver, self.balances[st_sender])
  File "/home/celioggr/erc20-pbt/erc20_pbt.py", line 41, in rule_transfer
    if self.DEBUG:
  File "/home/celioggr/erc20-pbt/env/lib/python3.8/site-packages/hypothesis/
      stateful.py", line 594, in rule_wrapper
    return f(*args, **kwargs)
  File "/home/celioggr/erc20-pbt/erc20_pbt.py", line 47, in rule_transfer
    tx = self.contract.transfer(
brownie.exceptions.VirtualMachineError: revert
Trace step -1, program counter 2896:
  File "contracts/INT.sol", line 135, in INTToken._transfer:
```

Listing B.77: INT3 - Falsifying example

```
Falsifying example:
state = BrownieStateMachine()
state.rule_transferAll(st_receiver=<Account '0
    x66aB6D9362d4F35596279692F0251Db635165871'>, st_sender=<Account '0
    x66aB6D9362d4F35596279692F0251Db635165871'>)
state.teardown()
```

B.5.1.4   INT4

This bug is the same as INT3 (B.5.1.3) but for transferFrom(). The resulting assertion error and
falsifying example are detailed in Listing B.78 and B.79 respectively.

Listing B.78: INT4 - Assertion error

```
Traceback (most recent call last):
  File "/home/celioggr/erc20-pbt/erc20_pbt.py", line 76, in rule_transferFrom
    tx = self.contract.transferFrom(
brownie.exceptions.VirtualMachineError: revert
Trace step -1, program counter 1476:
  File "contracts/INT.sol", line 67, in token.transferFrom:
```

Listing B.79: INT4 - Falsifying example

```
Falsifying example:
state = BrownieStateMachine()
state.rule_transferFrom(st_amount=10843, st_owner=<Account '0
    xA868bC7c1AF08B8831795FAC946025557369F69C'>, st_receiver=<Account '0
    x0063046686E46Dc6F15918b61AE2B121458534a5'>, st_spender=<Account '0
    x844ec86426F076647A5362706a04570A5965473B'>)
state.rule_transferFrom(st_amount=253, st_owner=<Account '0
    x66aB6D9362d4F35596279692F0251Db635165871'>, st_receiver=<Account '0
    x1CEE82EEd89Bd5Be5bf2507a92a755dcF1D8e8dc'>, st_spender=<Account '0
    x66aB6D9362d4F35596279692F0251Db635165871'>)
state.rule_transferFrom(st_amount=0, st_owner=<Account '0
    x23BB2Bb6c340D4C91cAa478EdF6593fC5c4a6d4B'>, st_receiver=<Account '0
    x46C0a5326E643E4f71D3149d50B48216e174Ae84'>, st_spender=<Account '0
    x33A4622B82D4c04a53e170c638B944ce27cffce3'>)
state.teardown()
```

B.5.1.5 INT5

In Listing B.80 line 2, is possible to see that an account can't transfer all allowance tokens. The resulting assertion error and falsifying example are detailed in Listing B.81 and B.82 respectively.

Listing B.80: InternetNodeToken - transferFrom() source code

```
1   function transferFrom(address _from, address _to, uint256 _value) returns (
        bool success) {
2       require (_value < allowance[_from][msg.sender]);      // Check allowance
3       allowance[_from][msg.sender] -= _value;
4       _transfer(_from, _to, _value);
5       return true;
6   }
```

Listing B.81: INT5 - Assertion error

```
Traceback (most recent call last):
  File "/home/celioggr/erc20-pbt/erc20_pbt.py", line 114, in
      rule_approveAndTransferAll
    self.rule_approve(st_owner, st_spender, amount)
  File "/home/celioggr/erc20-pbt/erc20_pbt.py", line 95, in rule_approve
    if self.DEBUG:
  File "/home/celioggr/erc20-pbt/env/lib/python3.8/site-packages/hypothesis/
      stateful.py", line 594, in rule_wrapper
    return f(*args, **kwargs)
  File "/home/celioggr/erc20-pbt/erc20_pbt.py", line 102, in rule_approve
    self.verifyEvent(
AssertionError: Approval: event was not fired
```

Listing B.82: INT5 - Falsifying example

```
Falsifying example:
state = BrownieStateMachine()
state.rule_approveAndTransferAll(st_owner=<Account '0
    x844ec86426F076647A5362706a04570A5965473B'>, st_receiver=<Account '0
    x844ec86426F076647A5362706a04570A5965473B'>, st_spender=<Account '0
    x844ec86426F076647A5362706a04570A5965473B'>)
state.teardown()
```

B.5.2 OpenZeppelin unit testing

Bug ID	Function	Assertion failure	Info	test ID
OZ_INT1	transfer/transferFrom	Operation not allowed	revert	1,2,5,6,7
OZ_INT2	transfer	Operation not allowed	transfer 0 tokens	3,4
OZ_INT3	approve	Absent event	missing Approval	8,9

B.5.2.1 OZ_INT1

The test IDs concerning bug ID **OZ_INT1** are detailed in Listings B.83, B.84, B.85, B.86 and B.87. This bug is related with bug ID INT3 (B.5.1.3).

Listing B.83: OZ_INT1 - test ID 1

```
1) Contract: INT
      transfer
        when the recipient is not the zero address
          when the sender transfers all balance
            transfers the requested amount:
    Error: Returned error: VM Exception while processing transaction: revert
```

Listing B.84: OZ_INT1 - test ID 2

```
2) Contract: INT
      transfer
        when the recipient is not the zero address
          when the sender transfers all balance
            emits a transfer event:
```

Listing B.85: OZ_INT1 - test ID 5

```
5) Contract: INT
      transfer from
        when the token owner is not the zero address
          when the recipient is not the zero address
            when the spender has enough approved balance
              when the token owner has enough balance
                transfers the requested amount:
    Error: Returned error: VM Exception while processing transaction: revert
```

Listing B.86: OZ_INT1 - test ID 6

```
6) Contract: INT
   transfer from
     when the token owner is not the zero address
       when the recipient is not the zero address
         when the spender has enough approved balance
           when the token owner has enough balance
             decreases the spender allowance:
   Error: Returned error: VM Exception while processing transaction: revert
```

Listing B.87: OZ_INT1 - test ID 7

```
7) Contract: INT
   transfer from
     when the token owner is not the zero address
       when the recipient is not the zero address
         when the spender has enough approved balance
           when the token owner has enough balance
             emits a transfer event:
   Error: Returned error: VM Exception while processing transaction: revert
```

B.5.2.2   OZ_INT2

The test IDs for bug ID **OZ_INT2** are detailed in Listings B.88 and B.89. In Listing B.71 line 4 it is possible to see that **_value** must be greater than zero for the function to proceed, thus not allowing for transfers with zero tokens.

Listing B.88: OZ_INT2 - test ID 3

```
3) Contract: INT
     transfer
       when the recipient is not the zero address
         when the sender transfers zero tokens
           transfers the requested amount:
   Error: Returned error: VM Exception while processing transaction: revert
```

Listing B.89: OZ_INT2 - test ID 4

```
4) Contract: INT
      transfer
        when the recipient is not the zero address
          when the sender transfers zero tokens
            emits a transfer event:
      Error: Returned error: VM Exception while processing transaction: revert
```

B.5.2.3 OZ_INT3

The test IDs for bug ID **OZ_INT3** are detailed in Listings B.90 and B.91. This bug is related with bug ID **INT2** (B.5.1.2) where **approve()** function fails to fire the proper event.

Listing B.90: OZ_INT3 - test ID 8

```
8) Contract: INT
    approve
      when the spender is not the zero address
        when the sender has enough balance
          emits an approval event:
  No 'Approval' events found
  + expected - actual
```

Listing B.91: OZ_INT3 - test ID 9

```
9) Contract: INT
    approve
      when the spender is not the zero address
        when the sender does not have enough balance
          emits an approval event:
  No 'Approval' events found
  + expected - actual
```

B.5.3 Consensys unit testing

Bug ID	Function	Assertion failure	Info	test ID
CS_INT1	transfer	Operation not allowed	reverts when balance(owner) = amount	1
CS_INT2	transfer	Operation not allowed	Dos not allow valid transfer (0 tokens)	2,3
CS_INT3	approve	Absent event	missing Approval	4

B.5.3.1 CS_INT1

Bug ID **CS_INT1** is related with **OZ_INT3** (B.5.1.3). The test ID for this bug is detailed in
Listing B.92.

Listing B.92: CS_INT1 - test ID 1

```
1) Contract: Internet Node Token
     transfers: should transfer 10000 to accounts[1] with accounts[0] having
         10000:
   Transaction: 0
         x1ce9a7b80d80890835de6d7e0e1e705671230ec16e1f37ce811edc3a73c497c4
         exited with an error (status 0).
   Please check that the transaction:
   − satisfies all conditions set by Solidity 'require' statements.
   − does not trigger a Solidity 'revert' statement.
```

B.5.3.2 CS_INT2

Consensys also found that tranfers with 0 tokens are not allowed, this is reported as bug id
CS_INT2, which is related with **OZ_INT2** (B.5.2.2). The test IDs for this bug are detailed in
Listings B.93 and B.94.

Listing B.93: CS_INT2 - test ID 2

```
2) Contract: Internet Node Token
     transfers: should handle zero−transfers normally:
   Transaction: 0
       x0fc6e7688aa97d38e810d67ef91c93ad87abe32f9791b2a44ac34d8a8fba0b35 exited
       with an error (status 0)
```

Listing B.94: CS_INT2 - test ID 3

```
3) Contract: Internet Node Token
     events: should fire Transfer event normally on a zero transfer:
   Transaction: 0
       xabbd340f9d8948fa576852aadfdd09b792245e384b042c1d9a00254f5e3c10ea exited
       with an error (status 0)
```

B.5.3.3 CS_INT3

Consensys reported one assertion failure related with the missing fire event Approve. This is
identified as **CS_INT3** and is related with **INT2** (B.5.1.2) and **OZ_INT3** (B.5.2.3). The test ID

for this bug is detailed in Listing B.95.

Listing B.95: CS_INT3 - test ID 4

```
4) Internet Node Token
      events: should fire Approval event properly:
      TypeError: Cannot read property 'args' of undefined
```

B.6 LinkToken

LinkToken only has one bug related with transactions that should **revert** when invalid conditions
are met, instead the contract's implementation returns a Boolean value to handle such scenarios.
All testing agents reported the same bugs.

B.6.1 Property-based testing

Bug ID	Rule	Assertion failure	Info
LINK1	transfer	Absent revert	Assertion error instead of reverting
LINK2	transferFrom	Absent revert	Assertion error instead of reverting

B.6.1.1 LINK1

The contract's library **SafeMath** (Listing B.96 line 4), makes bad use of **assert**() statement when
checking if transaction data is valid (line 4). For a more comprehensive description of this refer
to bug ID **USDT2** (B.8.1.2).

Listing B.96: Link SafeMath.sub() source code

```
1 library SafeMath {
2 // (...)
3 function sub(uint256 a, uint256 b) internal constant returns (uint256) {
4     assert(b <= a);
5     return a - b;
6   }
7 }
```

The assertion error and the correspondent falsifying example are illustrated in Listing B.97
in Listing B.98 for bug ID **LINK1**.

Listing B.97: LINK1 - Assertion example

```
Traceback (most recent call last):
  File "/home/celioggr/erc20-pbt/erc20_pbt.py", line 59, in rule_transfer
    self.contract.transfer(
brownie.exceptions.VirtualMachineError: invalid opcode: invalid opcode
```

Listing B.98: LINK1 - Falsifying example

```
Falsifying example:
state = BrownieStateMachine()
state.rule_transfer(st_amount=123, st_receiver=<Account '0
    x33A4622B82D4c04a53e170c638B944ce27cffce3'>, st_sender=<Account '0
    x21b42413bA931038f35e7A5224FaDb065d297Ba3'>)
state.teardown()
```

B.6.1.2 LINK2

Same as bug ID **LINK1** (B.6.1.1) but for **transferFrom**(). The assertion error and the correspondent falsifying example are illustrated in Listing B.99 and B.100 for bug ID **LINK2**.

Listing B.99: LINK2 - Assertion example

```
Traceback (most recent call last):
  File "/home/celioggr/erc20-pbt/erc20_pbt.py", line 90, in rule_transferFrom
    self.contract.transferFrom(
brownie.exceptions.VirtualMachineError: invalid opcode: invalid opcode
```

Listing B.100: LINK2 - Falsifying example

```
Falsifying example:
state = BrownieStateMachine()
state.rule_transferFrom(st_amount=256, st_owner=<Account '0
    x33A4622B82D4c04a53e170c638B944ce27cffce3'>, st_receiver=<Account '0
    x33A4622B82D4c04a53e170c638B944ce27cffce3'>, st_spender=<Account '0
    x66aB6D9362d4F35596279692F0251Db635165871'>)
state.teardown()
```

B.6.2   OpenZeppelin unit testing

Bug ID	Function	Assertion failure	Info	test ID
OZ_LINK1	transfer/transferFrom	Absent revert	did not revert (assert())	1,2,3,4

B.6.2.1   OZ_LINK1

This bug is related with bug ID LINK1 (B.6.1.1) and LINK2 (B.6.1.2). The test IDs concerning with bug ID OZ_LINK1 are detailed in Listings B.101, B.102, B.103 and B.104.

Listing B.101: OZ_LINK1 - test ID 1

```
1) Contract: LinkToken
   transfer
     when the recipient is not the zero address
       when the sender does not have enough balance
         reverts:
 Wrong kind of exception received
 + expected - actual
 -invalid opcode
 +revert
```

Listing B.102: OZ_LINK1 - test ID 2

```
2) Contract: LinkToken
   transfer from
     when the token owner is not the zero address
       when the recipient is not the zero address
         when the spender has enough approved balance
           when the token owner does not have enough balance
             reverts:
 Wrong kind of exception received
 + expected - actual
 -invalid opcode
 +revert
```

Listing B.103: OZ_LINK1 - test ID 3

```
3) Contract: LinkToken
   transfer from
     when the token owner is not the zero address
       when the recipient is not the zero address
         when the spender does not have enough approved balance
           when the token owner has enough balance
             reverts:
 Wrong kind of exception received
 + expected - actual
 -invalid opcode
 +revert
```

Listing B.104: OZ_LINK1 - test ID 4

```
4) Contract: LinkToken
   transfer from
     when the token owner is not the zero address
       when the recipient is not the zero address
         when the spender does not have enough approved balance
           when the token owner does not have enough balance
             reverts:
 Wrong kind of exception received
 + expected - actual
 -invalid opcode
 +revert
```

B.6.3 Consensys unit testing

Bug ID	Function	Assertion failure	Info	test ID
CS_LINK1	transfer/transferFrom	Absent revert	did not revert (assert())	1,2,3,4

B.6.3.1 CS_LINK1

This bug is related with bug ID **LINK1** (B.6.1.1) and **LINK2** (B.6.1.2). The test IDs concerning bug ID **CS_LINK1** are detailed in Listings B.105, B.106, B.107 and B.108.

Listing B.105: CS_LINK1 - test ID 1

```
1) Contract: ChainLink Token
        transfers: should fail when trying to transfer 10001 to accounts[1]
           with accounts[0] having 10000:
      AssertionError: Expected "revert", got StatusError: Transaction: 0
         xf3b413eecc013dc4eb67639e587f4b62cbe1800c0b5a578670e9d15604d5a093
         exited with an error (status 0) after consuming all gas.
      Please check that the transaction:
      - satisfies all conditions set by Solidity 'assert' statements.
      - has enough gas to execute the full transaction.
      - does not trigger an invalid opcode by other means (ex: accessing an
         array out of bounds). instead
```

Listing B.106: CS_LINK1 - test ID 2

```
2) Contract: ChainLink Token
     approvals: msg.sender approves accounts[1] of 100 & withdraws 50 & 60 (2nd
        tx should fail):
   AssertionError: Expected "revert", got StatusError: Transaction: 0
      x898c3db01c1037f11cf436119c9ced03a8300b005c93ff42b5bafd997484667d exited
      with an error (status 0) after consuming all gas.
   Please check that the transaction:
   - satisfies all conditions set by Solidity 'assert' statements.
   - has enough gas to execute the full transaction.
   - does not trigger an invalid opcode by other means (ex: accessing an array
      out of bounds). instead
```

Listing B.107: CS_LINK1 - test ID 3

```
3) Contract: ChainLink Token
     approvals: attempt withdrawal from account with no allowance (should fail):
   AssertionError: Expected "revert", got StatusError: Transaction: 0
      xbcfbe0624a05e414af76cd83c05abe0183d0f7da42ab940d8163c9bdbab6bc83 exited
      with an error (status 0) after consuming all gas.
   Please check that the transaction:
   - satisfies all conditions set by Solidity 'assert' statements.
   - has enough gas to execute the full transaction.
   - does not trigger an invalid opcode by other means (ex: accessing an array
      out of bounds). instead
```

Listing B.108: CS_LINK1 - test ID 4

```
4) Contract: ChainLink Token
    approvals: allow accounts[1] 100 to withdraw from accounts[0]. Withdraw 60
        and then approve 0 & attempt transfer.:
  AssertionError: Expected "revert", got StatusError: Transaction: 0
      xde63277171858c1170d50f3a40ae89125b40f569117d6204b5e1d1591c2ca0ff exited
      with an error (status 0) after consuming all gas.
  Please check that the transaction:
  - satisfies all conditions set by Solidity 'assert' statements.
  - has enough gas to execute the full transaction.
  - does not trigger an invalid opcode by other means (ex: accessing an array
      out of bounds). instead
```

B.7 SwftCoin

This token contract shares most of its code (including **approve()** implementation) with BitAsean-Token (Section B.1). Thus, the same bug related with the missing fire event **Approval** will persist in this analysis **BAS1** (B.1.1.1). This bug was reported by all agents. The second bug is related with an absent return value for a valid transfer that was reported only by Brownie.

B.7.1 Property-based testing

Bug ID	Rule	Assertion failure	Info
SWFTC1	approve	Absent event	Approval event not fired for valid approval
SWFTC2	transfer	Absent return value	Fails to return a value for valid transfer

B.7.1.1 SWFTC1

In Listing B.109, we see that contract's implementation of **approve()** does not fire **Approval** event. The assertion error and the correspondent falsifying example are illustrated in Listing B.110 and B.111.

Listing B.109: SwftCoin approve() implementation

```
function approve(address _spender, uint256 _value)
  returns (bool success) {
  allowance[msg.sender][_spender] = _value;
  return true;
}
```

Listing B.110: SWFTC1 - Assertion error

```
def verifyEvent(self, tx, eventName, data):
    if not eventName in tx.events:
>           raise AssertionError("{}: event was not fired".format(eventName))
E           AssertionError: Approval: event was not fired
```

Listing B.111: SWFTC1 - Falsifying example

```
Falsifying example:
state = BrownieStateMachine()
state.rule_transfer(st_amount=1, st_receiver=<Account '0
    x66aB6D9362d4F35596279692F0251Db635165871'>, st_sender=<Account '0
    x66aB6D9362d4F35596279692F0251Db635165871'>)
state.teardown()
```

B.7.1.2 SWFTC2

This bug is related with **BAS2** (B.1.1.2). In Listing B.112, is possible to see that **transfer** () does not return **true** for a valid transfer. The resulting assertion error and falsifying example are detailed in Listing B.113 and B.114 respectively.

Listing B.112: SwftCoin - **transfer** () source code

```
function transfer(address _to, uint256 _value) {
    if (balanceOf[msg.sender] < _value) throw;
    if (balanceOf[_to] + _value < balanceOf[_to]) throw;
    balanceOf[msg.sender] -= _value;
    balanceOf[_to] += _value;
    Transfer(msg.sender, _to, _value);
}
```

Listing B.113: SWFTC2 - Assertion error

```
Traceback (most recent call last):
  File "/home/celioggr/erc20-pbt/erc20_pbt.py", line 43, in rule_transfer
    self.verifyReturnValue(tx, True)
  File "/home/celioggr/erc20-pbt/erc20_pbt.py", line 109, in verifyReturnValue
    self.verifyValue("return value", expected, tx.return_value)
  File "/home/celioggr/erc20-pbt/erc20_pbt.py", line 114, in verifyValue
    raise AssertionError("{} : expected value {}, actual value was {}".format(
        msg, expected, actual))
AssertionError: return value : expected value True, actual value was None
```

Listing B.114: SWFTC2 - Falsifying example

```
Falsifying example:
state = BrownieStateMachine()
state.rule_transfer(st_amount=1, st_receiver=<Account '0
    x0063046686E46Dc6F15918b61AE2B121458534a5'>, st_sender=<Account '0
    x66aB6D9362d4F35596279692F0251Db635165871'>)
state.teardown()
```

B.7.2 OpenZeppelin unit testing

Bug ID	Function	Assertion failure	Info	test ID
OZ_SWFTC1	approve	Absent event	Approval event not fired for valid	1,2

B.7.2.1 OZ_SWFTC1

The test IDs for bug ID **OZ_SWFTC1** are detailed in Listings B.115 and B.116. This bug is related with bug ID **SWFTC1** (B.7.1.1).

Listing B.115: OZ_SWFTC1 - test ID 1

```
1) Contract: SwftCoin
     approve
       when the spender is not the zero address
         when the sender has enough balance
           emits an approval event:

  No 'Approval' events found
  + expected - actual
```

Listing B.116: OZ_SWFTC1 - test ID 2

```
2) Contract: SwftCoin
       approve
         when the spender is not the zero address
           when the sender does not have enough balance
             emits an approval event:

    No 'Approval' events found
    + expected - actual
```

B.7.3 Consensys unit testing

Bug ID	Function	Assertion failure	Info	test ID
CS_SWFTC1	approve	Absent event	Approval event not fired for valid	1

B.7.3.1 CS_SWFTC1

The test ID for bug ID **OZ_SWFTC1** are detailed in Listing B.117. This bug is related with bug ID **SWFTC1** (B.7.1.1).

Listing B.117: CS_SWFTC1 - test ID 1

```
1) Contract: SwftCoin
     events: should fire Approval event properly:
  TypeError: Cannot read property 'args' of undefined
```

B.8 TetherToken

This contract counts with 6 bugs. All bugs were reported by the testing agents apart from two of them related with absent values for valid transactions. These were not reported by the unit testing agents. Consensys failed to detect the approve mitigation attack bug, since it does not attempt to replay approve calls. Assuming that, if the first approve call has succeeded, then there is no need for further testing.

B.8.1 Property-based testing

Bug ID	Rule	Assertion failure	Info
USDT1	approve	Operation not allowed	race condition mitigation
USDT2	approve	Absent return value	fails to return value for valid approve
USDT3	transfer	Absent revert	Assertion error instead of revert
USDT4	transfer	Absent return value	Fails to return value for valid approve
USDT5	transferFrom	Absent revert	Assertion error instead of revert
USDT6	transferFrom	Absent return value	Fails to return value for valid transfer

B.8.1.1 USDT1

This bug is related with a race condition mitigation implemented on **approve()** [57]. Although the standard does not state how this attack vector should be addressed, it refers to it in Listing

B.119. In Listing B.118 line 3, we can see that, if there is an allowance set then updates to this allowance must reset its value to zero before new approvals.

Listing B.118: TetherToken - approve() source code

```
1    function approve(address _spender, uint _value)
2    public onlyPayloadSize(2 * 32) {
3      require(!((_value != 0) && (allowed[msg.sender][_spender] != 0)));
4      allowed[msg.sender][_spender] = _value;
5      Approval(msg.sender, _spender, _value);
6    }
```

Listing B.119: EIP-20:ERC-20 - approve()

```
NOTE: To prevent attack vectors like the one described here and discussed here
    , clients SHOULD make sure to create user interfaces in such a way that
    they set the allowance first to 0 before setting it to another value for
    the same spender. THOUGH The contract itself shouldn't enforce it, to
    allow backwards compatibility with contracts deployed before
```

The attack vector can be described as the following scenario.

1. Bob approves Alice to spend 10 tokens on his behalf.

   approve(Alice ,10,{' from ': Bob)}

2. However, Bob realizes that 5 tokens was the amount he wanted to allow to Alice in the first place. He submits an update to the previous allowance.

   approve(Alice ,5,{' from ': Bob)}

Although, seconds before step 2, Alice already transferred 10 tokens on Bob's behalf to another account. This transaction was mined by the time Bob realized his mistake and issued an allowance update. After the allowance update issued by Bob, Alice will be able to withdraw 5 more tokens. value.

The assertion error and the correspondent falsifying example are illustrated in Listing B.120 and B.121 for bug ID **USDT1**.

Listing B.120: USDT1 - Assertion error

```
Traceback (most recent call last):
  File "/home/celioggr/erc20-pbt/erc20_pbt.py", line 98, in rule_approve
    tx = self.contract.approve(
brownie.exceptions.VirtualMachineError: revert
Trace step -1, program counter 4573:
```

Listing B.121: USDT1 - Falsifying example

```
Falsifying example:
state = BrownieStateMachine()
state.rule_approve(st_amount=1024, st_owner=<Account '0
    x33A4622B82D4c04a53e170c638B944ce27cffce3'>, st_spender=<Account '0
    x33A4622B82D4c04a53e170c638B944ce27cffce3'>)
state.rule_approve(st_amount=1024, st_owner=<Account '0
    x33A4622B82D4c04a53e170c638B944ce27cffce3'>, st_spender=<Account '0
    x33A4622B82D4c04a53e170c638B944ce27cffce3'>)
state.teardown()
```

B.8.1.2 USDT2

This bug is related with an absent return value for a valid approval. In Listing B.118 we can
see that for valid approvals the function fails to return **true** as according to the standard. The
assertion error and the correspondent falsifying example are illustrated in Listing B.122 and
B.123 for bug ID USDT2.

Listing B.122: USDT2 - Assertion error

```
Traceback (most recent call last):
  File "/home/celioggr/erc20-pbt/erc20_pbt.py", line 114, in
      rule_approveAndTransferAll
    self.rule_approve(st_owner, st_spender, amount)
  File "/home/celioggr/erc20-pbt/erc20_pbt.py", line 95, in rule_approve
    if self.DEBUG:
  File "/home/celioggr/.local/lib/python3.8/site-packages/hypothesis/stateful.
      py", line 594, in rule_wrapper
    return f(*args, **kwargs)
  File "/home/celioggr/erc20-pbt/erc20_pbt.py", line 107, in rule_approve
    self.verifyReturnValue(tx, True)
AssertionError: return value : expected value True, actual value was None
```

Listing B.123: USDT2 - Falsifying example

```
Falsifying example:
state = BrownieStateMachine()
state.rule_approveAndTransferAll(st_owner=<Account '0
    x33A4622B82D4c04a53e170c638B944ce27cffce3'>, st_receiver=<Account '0
    x33A4622B82D4c04a53e170c638B944ce27cffce3'>, st_spender=<Account '0
    x66aB6D9362d4F35596279692F0251Db635165871'>)
state.teardown()
```

B.8.1.3   USDT3

The contract makes improper use of **assert**() function when executing math operations with safety checks (Listing B.124, line 3). When dealing with input validation the recommendation is to use **require**() statements. In scenarios where invalid transactions are issued, the standard states that the call should **throw**. The use of **assert**() returns an invalid opcode assertion error (Listing B.125, line 4).

The assertion error and the correspondent falsifying example are illustrated in Listing B.125 and B.126 for bug ID **USDT3**.

Listing B.124: TetherToken - SafeMath.sub()

```
1    function sub(uint256 a, uint256 b)
2    internal pure returns (uint256) {
3      assert(b <= a);
4      return a - b;
5  }
```

Listing B.125: USDT3 - Assertion error

```
Traceback (most recent call last):
  File "/home/celioggr/erc20-pbt/erc20_pbt.py", line 59, in rule_transfer
    self.contract.transfer(
brownie.exceptions.VirtualMachineError: invalid opcode
Trace step -1, program counter 5704:
  File "contracts/TetherToken.sol", line 29, in SafeMath.sub:
        }
        function sub(uint256 a, uint256 b) internal pure returns (uint256) {
            assert(b <= a);
            return a - b;
        }
```

Listing B.126: USDT3 - Falsifying example

```
Falsifying example:
state = BrownieStateMachine()
state.rule_transfer(st_amount=1, st_receiver=<Account '0
    x33A4622B82D4c04a53e170c638B944ce27cffce3'>, st_sender=<Account '0
    x33A4622B82D4c04a53e170c638B944ce27cffce3'>)
state.teardown()
```

B.8.1.4   USDT4

This bug is related with an absent return value for a valid transfer. In Listing B.127 we can see
that for valid transfers the function fails to return **true**. The assertion error and the correspondent
falsifying example are illustrated in Listing B.128 and B.129 for bug ID **USDT4**.

Listing B.127: TetherToken - transfer() source code

```
function transfer(address _to, uint _value) public onlyPayloadSize(2 * 32) {
  uint fee = (_value.mul(basisPointsRate)).div(10000);
  if (fee > maximumFee) {
      fee = maximumFee;
  }
  uint sendAmount = _value.sub(fee);
  balances[msg.sender] = balances[msg.sender].sub(_value);
  (...)
  Transfer(msg.sender, _to, sendAmount);
}
```

Listing B.128: USDT4 - Assertion error

```
Traceback (most recent call last):
  File "/home/celioggr/erc20-pbt/erc20_pbt.py", line 110, in rule_transferAll
    self.rule_transfer(st_sender, st_receiver, self.balances[st_sender])
  File "/home/celioggr/erc20-pbt/erc20_pbt.py", line 41, in rule_transfer
    if self.DEBUG:
  File "/home/celioggr/.local/lib/python3.8/site-packages/hypothesis/stateful.
    py", line 594, in rule_wrapper
    return f(*args, **kwargs)
  File "/home/celioggr/erc20-pbt/erc20_pbt.py", line 56, in rule_transfer
    self.verifyReturnValue(tx, True)
AssertionError: return value : expected value True, actual value was None
```

Listing B.129: USDT4 - Falsifying example

```
Falsifying example:
state = BrownieStateMachine()
state.rule_transferAll(st_receiver=<Account '0
    x66aB6D9362d4F35596279692F0251Db635165871'>, st_sender=<Account '0
    x66aB6D9362d4F35596279692F0251Db635165871'>)
state.teardown()
```

B.8.1.5 USDT5

This bug is the same as USDT3 (B.8.1.3) but for transferFrom(). The assertion error and the correspondent falsifying example are illustrated in Listing B.130 and B.131 for bug ID USDT5.

Listing B.130: USDT5 - Assertion error

```
Traceback (most recent call last):
  File "/home/celioggr/erc20-pbt/erc20_pbt.py", line 90, in rule_transferFrom
    self.contract.transferFrom(
brownie.exceptions.VirtualMachineError: invalid opcode: dev: assert opcode
Trace step -1, program counter 5704:
  File "contracts/TetherToken.sol", line 29, in SafeMath.sub:
        }
        function sub(uint256 a, uint256 b) internal pure returns (uint256) {
            assert(b <= a); // dev: assert opcode
            return a - b;
        }
```

Listing B.131: USDT5 - Falsifying example

```
Falsifying example:
state = BrownieStateMachine()
state.rule_transferFrom(st_amount=75, st_owner=<Account '0
    x66aB6D9362d4F35596279692F0251Db635165871'>, st_receiver=<Account '0
    x0063046686E46Dc6F15918b61AE2B121458534a5'>, st_spender=<Account '0
    x33A4622B82D4c04a53e170c638B944ce27cffce3'>)
state.teardown()
```

B.8.1.6 USDT6

This bug is related with an absent return value for a valid transfer and related with bug ID USDT4 (B.8.1.4) but for transferFrom(). In Listing B.127, we can see that for valid transfers the function fails to return true as according to the standard. The assertion error and the correspondent falsifying example are illustrated in Listing B.132 and B.133 for bug ID **USDT6**.

Listing B.132: USDT6 - Assertion error

```
Traceback (most recent call last):
  File "/home/celioggr/erc20-pbt/erc20_pbt.py", line 87, in rule_transferFrom
    self.verifyReturnValue(tx, True)
AssertionError: return value : expected value True, actual value was None
```

Listing B.133: USDT6 - Falsifying example

```
Falsifying example:
state = BrownieStateMachine()
state.rule_transferFrom(st_amount=0, st_owner=<Account '0
    x0063046686E46Dc6F15918b61AE2B121458534a5'>, st_receiver=<Account '0
    x0063046686E46Dc6F15918b61AE2B121458534a5'>, st_spender=<Account '0
    x66aB6D9362d4F35596279692F0251Db635165871'>)
state.teardown()
```

B.8.2   OpenZeppelin unit testing

Bug ID	Function	Assertion failure	Info	test ID
OZ_USDT1	approve	Operation not allowed	race condition mitigation	5,6
OZ_USDT2	transfer/transferFrom	Absent revert	sub() bad implementation	1,2,3,4

B.8.2.1   OZ_USDT1

The test IDs for bug ID **OZ_USDT1** are detailed in Listings B.134, B.135. This bug is related
with bug ID **USDT1** (B.8.1.1).

Listing B.134: OZ_USDT1 test ID 5

```
5) Contract: TetherToken
       approve
           when the spender is not the zero address
             when the sender has enough balance
               when the spender had an approved amount
                 approves the requested amount and replaces the previous one:
       Error: Returned error: VM Exception while processing transaction: revert
```

Listing B.135: OZ_USDT1 - test ID 6

```
6) Contract: TetherToken
      approve
        when the spender is not the zero address
          when the sender does not have enough balance
            when the spender had an approved amount
              approves the requested amount and replaces the previous one:
  Error: Returned error: VM Exception while processing transaction: revert
```

B.8.2.2   OZ_USDT2

The test IDs for bug ID **OZ_USDT2** are detailed in Listings B.136, B.137, B.138 and B.139. This bug is related with bug ID **USDT3** (B.8.1.3) and **USDT5** (B.8.1.5).

Listing B.136: OZ_USDT2 - test ID 1

```
1) Contract: TetherToken
   transfer
     when the recipient is not the zero address
       when the sender does not have enough balance
         reverts:
   Wrong kind of exception received
   + expected - actual
   -invalid opcode
   +revert
```

Listing B.137: OZ_USDT2 - test ID 2

```
2) Contract: TetherToken
   transfer from
     when the token owner is not the zero address
       when the recipient is not the zero address
         when the spender has enough approved balance
           when the token owner does not have enough balance
             reverts:
   Wrong kind of exception received
   + expected - actual
   -invalid opcode
   +revert
```

Listing B.138: OZ_USDT2 - test ID 3

```
3) Contract: TetherToken
   transfer from
     when the token owner is not the zero address
       when the recipient is not the zero address
         when the spender does not have enough approved balance
           when the token owner has enough balance
             reverts:
Wrong kind of exception received
+ expected − actual
 −invalid opcode
 +revert
```

Listing B.139: OZ_USDT2 - test ID 4

```
4) Contract: TetherToken
   transfer from
     when the token owner is not the zero address
       when the recipient is not the zero address
         when the spender does not have enough approved balance
           when the token owner does not have enough balance
             reverts:
Wrong kind of exception received
+ expected − actual
 −invalid opcode
 +revert
```

B.8.3 Consensys unit testing

Bug ID	Function	Assertion failure	Info	test ID
CS_USDT1	transfer/transferFrom	Absent revert	sub() bad implementation	1,2,3,4

B.8.3.1 CS_USDT1

The test IDs for bug ID **CS_USDT1** are detailed in Listings B.140, B.141, B.142 and B.143. This bug is related with bug ID **UDST3** (B.8.1.3) and **USDT5** (B.8.1.5).

Listing B.140: CS_USDT1 - test ID 1

```
1) Contract: TetherToken
      transfers: should fail when trying to transfer 10001 to accounts[1]
          with accounts[0] having 10000:
      AssertionError: Expected "revert", got StatusError: Transaction: 0
          xf3b413eecc013dc4eb67639e587f4b62cbe1800c0b5a578670e9d15604d5a093
          exited with an error (status 0) after consuming all gas.
      Please check that the transaction:
      - satisfies all conditions set by Solidity 'assert' statements.
      - has enough gas to execute the full transaction.
      - does not trigger an invalid opcode by other means (ex: accessing an
          array out of bounds). instead
```

Listing B.141: CS_USDT1 - test ID 2

```
2) Contract: TetherToken
      approvals: msg.sender approves accounts[1] of 100 & withdraws 50 & 60
          (2nd tx should fail):
      AssertionError: Expected "revert", got StatusError
```

Listing B.142: CS_USDT1 - test ID 3

```
3) Contract: TetherToken
    approvals: attempt withdrawal from account with no allowance (should fail):
  AssertionError: Expected "revert
```

Listing B.143: CS_USDT1 - test ID 4

```
4) Contract: TetherToken
      approvals: allow accounts[1] 100 to withdraw from accounts[0]. Withdraw
          60 and then approve 0 & attempt transfer.:
      AssertionError: Expected "revert"
```

9 783384 242624